BOTTLENECK

Bottleneck

Humanity's Impending Impasse

William R. Catton, Jr.

To order additional copies of this book, contact:
Xlibris Corporation
1-888-795-4274
www.Xlibris.com
Orders@Xlibris.com
60202

CONTENTS

To Nancy and our sons
in loving appreciation
for steadfastly insisting
this book needed to be written.

May future generations of people
inhabiting this planet be descended from
the most hubris-free members
of each preceding generation.

From 1776,
when the Newcommon steam engine
had been upgraded by James Watt,
its use led to escalating reliance
on fossil energy,
temporarily giving
increasing fractions
of the world's human population
gigantic powers.
With subsequent technological developments
Homo colossus acquired through
the next nine generations
the delusion of limitlessness.

Now this
from the "Controller" in the tower:

HUBRIS 1776 ABORT YOUR TAKEOFF!
I SAY AGAIN ABORT YOUR TAKEOFF IMMEDIATELY!
YOU ARE ATTEMPTING TO TAKE OFF FROM
A RUNWAY THAT IS TOO SHORT.
TAKEOFF CLEARANCE CANCELED.
ACKNOWLEDGE.

PREFACE

"Don't just stand there; *do* something!" That is a sentiment commonly felt by people confronting a severe challenge. But it sometimes happens that what gets done is counterproductive—makes matters worse not better, perhaps very much worse. The book you have begun to read is meant to warn that all of humanity is today confronted with a situation where the things we are too likely to do as we respond to challenging times will make our real but habitually misunderstood predicament very much worse.

A pattern of counterproductive response to a dire situation can arise from ignorance of real cause-effect relations (from false or illusory notions commonly held about how the world's processes actually work). Consider an example: the dying of the first president of the United States. George Washington was only 67 when he contracted a sudden illness that need not have been fatal. His death was *hastened by the adverse effects of then-conventional treatment for myriad types of illness.* Bleeding the patient was a standard medical procedure in those days, as it had been for many previous centuries, and remained so for several more decades. Having come down with a bad cold and a severe sore throat, Washington was bled several times in one day by his doctors, who cumulatively drained nearly half his body's total blood supply. Medical historians consider it probable that this "remedial" treatment killed him, not the condition to which it was the response.

In the 21st century we must beware of fatal "solutions" to misperceived calamities. Enormous troubles for all of humanity lie ahead. These will both result from and result in extraordinary social change. Tragically, insofar as people continue interpreting events according to an obsolete worldview, change will be too often ill-conceived and troubles will be seriously misunderstood—and thus worsened. Obsolete assumptions that developed in past centuries under circumstances now fundamentally changed remain too prevalent. There were then many fewer of us, we had much less powerful apparatus for living, and the planet's humanly usable resources had only begun to be ravenously exploited. In the present century prodigious efforts to cope with new challenges will be horribly counterproductive unless they are based upon new ways of thinking, mindful as never before of the human species' multifaceted involvement in the intricacies of the whole web of life on this planet. Tunnel vision must be overcome.

You may recall seeing a "news" item about "Tax Freedom Day," published each spring by an interest group called "Tax Foundation." Its intended implication is that all the money you earned until that far along in the year goes to pay the nation's annual tax "burden." Only from the next day onward, they want you to feel, is what you earn yours to spend as you choose. The date occurs late enough each year to impress readers with the Foundation's notion that an insufferably large and growing fraction of people's hard-earned income is being unjustly taken from them by a ghostly thief, "the government." But members of the reading public are also familiar with another politically poisonous phrase, "deficit spending," used to insinuate villainous extravagance by government for "spending more than it takes in" (by taxation). Any party out of power tends to fret about such deficits—until it wins election, and assumes power.

Ironically, these two concepts, which seem contrary to each other, can be combined. A person can deplore both a deficit *and* a tax burden. But opposite insights combined can yield enhanced illumination. Given some necessary level of civilization-maintaining collective expenditure, taxation is a necessary antidote to a deficit. One can learn to deplore the prior inability to recognize that fact.

Less commonly deplored, though, is a deficit due to the ravenous use not of money but of *natural resources* by modern civilization. We *could* each year mark a "carrying capacity day," the date (shockingly *early* in the year) by which all the energy and other resources consumed to that point on the calendar matches the total which the Earth's natural processes would require a whole year to *replenish*. And it ought to be possible to speak meaningfully of the "carrying capacity deficit" inflicted when the load imposed by humans upon the biosphere thus exceeds a sustainable magnitude.

The BBC's science and nature reporter was speaking along precisely those lines in April 2006 when he described a report by the New Economics Foundation and the Open University that the United Kingdom was "about to run out of its own natural resources and become dependent on supplies from abroad." The report gave 16 April as the day when that island country goes into "ecological debt." More important, it warned that if *worldwide* resource use levels matched those of modern Britain it would take more than three Earths to meet the demand.

As my 1980 book *Overshoot* was meant to show, those unfamiliar (but ecologically meaningful) concepts—carrying capacity, and carrying capacity deficit—are more helpful than the customary terms of ordinary political discourse in enabling people to understand how and why their world is changing and how their lives will be impacted in decades ahead. Now at last, people who think with a genuine ecological vocabulary do seem to be increasing in number, but still they are a small minority of even all literate people.

Perhaps that is as it must be. Organized human life would be impossible if we insisted *everyone's* profession should be the study of natural resources and the rate of their use. For life to go on we need people to work at *many* specialties. The cliché "to each his own" applies to the world of work, as well as to the arena of taste. To live more than a very primitive life, we need an intricate *division of labor*. It "takes all kinds of people" to make a functioning society.

In this book I will argue, however, that human societies are in deep trouble today from the converging influences of several trends left festering too long. First, there are many more of us inhabiting this planet than our forebears ever anticipated. On average, too, we are effectively much "larger" than our ancestors. By that I mean that the many technological advances of the past couple of centuries have greatly enlarged our per capita resource appetites and our per capita environmental impacts, as well as our capability to do harm to one another. Thirdly, we continue giving attention too much to what economists are fond of calling "the near term." Even though, as the symbol-using species, we do have somewhat more capability than other creatures to anticipate the future and plan ahead, old habits persist (partly for a reason we seldom consider, which I shall describe in this book) so our foresight has not been at all commensurate with our cumulative impact on the future conditions of this planet upon which our lives will continue to depend. So the twenty-first century holds perils vastly more serious than have been anticipated even by so-called "alarmists." We cannot afford misperceptions that obscure or understate and thereby unintentionally exacerbate the perils of this bottleneck century.

Some misperceptions probably arise from occupational specialization. When I was a graduate student in the 1950s, one of my sociology professors commented that becoming identified with any one academic discipline tends to free us from obligation to have extensive and detailed acquaintance with what is known in other disciplines. In short, he appreciated an emancipative effect of academic division of labor. I went on to a rewarding career in university teaching and research, probably benefitting all along from that emancipation, not needing to know more than peripherally the subject-matter accumulating in most disciplines other than my own. But eventually I came to deplore (as myopic) both sociology's separation from anthropology and its premature divorce from evolutionary biology. And for me, retirement has been fun because it was liberating in the opposite way. I found myself free at last to spend my time reading whatever interested me, rather than what I needed to cover in next week's lectures. And so I had the wonderfully human pleasure of widening my horizons as I continued to learn new things.

Specialists could be more efficient than jacks-of-all-trades, said the early economic theorist Adam Smith. In *Wealth of Nations* (1776) he claimed greatly increased productivity is the big advantage of division of labor. My graduate school professor was just insisting the same advantage holds for the production of knowledge as truly as it applied to collaboration among different specialists in the famous "pin factory" example by which Smith illustrated the idea.

The pioneering French sociologist Emile Durkheim (1893) saw a different advantage. Since specialists cannot by definition be self-sufficient, their necessary exchange relations make them interdependent. Interdependence would be, Durkheim supposed, the basis for a new form of social cohesion. It would gradually replace the former solidarity based on cultural homogeneity, already eroding in his time.

In contrast, his American contemporary, E. A. Ross, in a little book entitled *Sin and Society* (1907), argued that as technological advances facilitate further division of labor, cohesive social life is jeopardized by the increasing ways in which people can harm others (inadvertently, as well as sometimes deliberately). Division of labor dilutes responsibility.

My view today is that Durkheim was indeed spotlighting a fundamental fact about people's lives when he studied division of labor. But he didn't recognize all of what the spotlight revealed. The continued further branching of human society's "occupational bush" has certainly been a basic feature of the history of recent centuries. Growing ever bushier, the structure of modern living has made us ultimately dependent upon a fantastically intricate web of exchange relations called "the market." What Durkheim did not foresee, though, is that this trend was going to have a serious and *pervasive* dehumanizing effect.

That was an additional insight Ross was pointing to but did not fully see either. Experiences of my lifetime, together with much of my reading in recent years, led me to the view that the Durkheim insight and the Ross insight must be joined to enable us to understand modern history, what we have done to our world, and what will consequently be happening to us. Durkheim was overly sanguine that the effects of increasing interdependence among occupationally differentiated people would be beneficial. Ross was too sanguine about expecting humans, if enlightened about their increasing vulnerability, to have the will and ability to resist exploitation, perhaps by simply "throwing the rascals out."

To see what's ahead more clearly, let us combine their major insights, and view the combination in terms of what has been learned about evolutionary processes and ecological realities. We must take into account the trio of disturbing trends mentioned above (humans having become so numerous, so ravenous, and so short-sighted). All these cast a very different light on the human prospect. In modern overgrown and overdeveloped societies, whatever their official ideology may be, other humans (apart from those in our own circle of friends and relations) come to be seen too largely as mere "resources"—either as providers of products and services we covet, or as "walking billfolds" from whom we strive to extract the means to purchase products and services.

The interdependence generated by division of labor has made money an essential aspect of life. This drives people toward seeing "the economy" as a money tree. We grow up learning ways to participate in plucking money tree foliage. As distinctions between fair means and foul tend to fade, "social" and "humanitarian" concerns become just "special interests" rather than a universal ethic. Some of the very attributes that have enabled the species *Homo sapiens* to thrive, proliferate, and achieve great progress, are now producing these self-destructive tendencies.

The world's people are still divided by racial, nationality, and religious distinctions and any of these can be turned into antagonisms. And now we must also recognize the dehumanizing influence of today's elaborate division of labor.

Has this become mankind's unaffordable last divisive straw?

CHAPTER 1

Divisive Modernity

If one counts only the size of houses and cars, and the numbers of electronic gadgets stuffed into rec rooms, Americans are probably better off than ever before. But . . . well-being doesn't come just from piling up toys. An economy has psychological or, if you will, spiritual, dimensions. A conviction of fairness, a feeling of not being totally on one's own, a sense of reasonable stability and predictability are all essential . . . .

—Charles R. Morris 2006 "Freakoutonomics."
OpEd, *The New York Times*, June 2.

People differ.

Some differences are innate, some develop as we go through life. We have many ethnicities, many levels of education, many religious faiths, various political persuasions. We vary in affluence, from grandiose wealth to bitter poverty—either of which can be dehumanizing. We come in various sizes and shapes, and two sexes. In any human community, there are usually some old people, some middle-aged people, some young people, and some infants. Different sorts of activities are exhibited by (and expected of) individuals of different ages. There are still role-differences by gender, although our expectations on this have changed and continue to change. Sociology's most distinctive task, it has been suggested, was analysis of the various forms of differentiation, how these interrelate, what conditions produce them, how they become altered, and what they imply for societal life.[1]

In modern societies there are extreme instances of occupational specialization. The diversity of occupations has become enormous. No member of a modern human society can get through life without dependence on others. The division of labor in human societies has been a classic topic in sociology. The traditional view among sociologists was: lose self-sufficiency, gain cohesion. Supposedly, by making us highly interdependent, our division of labor has wrought a strong new form of social solidarity.[2]

My aim in writing this book has been to challenge that view. Both from my studies, and from decades of experience, I see abundant indications that modern division of labor functions too prevalently as a *divisive* factor, *eroding* rather than fostering social solidarity. People today may indeed be unable to be anywhere near self-*sufficient*, but the very forces making that so tend to push us toward being self-*centered*—often confused about it and vaguely frustrated. Every day the news includes innumerable symptoms of endemic self-centeredness. My aim is to show how modern division of labor fosters that condition. It is a condition that may be humanity's undoing. Nothing in nature guarantees the permanence of any particular human society, nor even of the human species.

Pervasive Anxiety Eroding Solidarity

Eugene Robinson, *Washington Post* columnist, wrote in May, 2006, that he found it "unnerving to see the country so unnerved," referring to the astonishing *lack* of the "thundering outrage from sea to shining sea" he would have expected in response to news that the federal government was trying to analyze cumulative records of who telephones whom in America. He suggested acquiescent apathy about "intrusive domestic surveillance" was due to the normally sunny optimism of Americans having yielded to emotions we handle poorly—"fear, insecurity, resentment." In previous columns, Robinson had accused the president (George W. Bush) of stoking fears and regularly exploiting them as a means of getting Americans to accept the previously unacceptable. However, he acknowledged that the administration had not conjured from thin air this apprehensive mood among the American people. Its roots went deeper.

By invoking a metaphoric caricature of the nation reclining as a patient on a therapist's couch, Robinson said the psychiatrist's notes at the end of the session might read as follows:

> Patient feels vulnerable to attack; cannot remember having experienced similar feeling before. Patient accustomed to being in control; now feels buffeted by outside forces beyond grasp. Patient believes livelihood and prosperity being usurped by others (repeatedly mentions China). Patient seeks scapegoats for personal failings (immigrants, Muslims,

civil libertarians). Patient is by far most powerful nation in world, yet feels powerless. Patient is full of unfocused anger.

There could hardly be a more vivid contrast than the difference so immediately evident between that mood in 2006 and the one induced seventy-three years earlier by the deliberately reassuring words of an earlier president, spoken upon his taking office when the country was in the depths of a great economic depression. Then too, there was anger among the American public. In those days some of the anger was perhaps too often too easily focused on the previous White House occupant, Herbert Hoover—for it was easier to blame him than to consider the fact that the economic hard times were international and not uniquely American. They were not simply a result of the putative inadequacy of one nation's unlucky head of government. President Hoover's successor, Franklin Delano Roosevelt, after taking the oath of office on March 4th, 1933, had said it was preeminently "a time to speak the truth, the whole truth, frankly and boldly. Nor need we shrink," he said,

> from honestly facing conditions in our country today. This great Nation will endure as it has endured, will revive and will prosper. So, first of all, let me assert my firm belief that the only thing we have to fear is fear itself—*nameless, unreasoning, unjustified terror which paralyzes needed efforts to convert retreat into advance.*

I have italicized Roosevelt's additional qualifying words, for they are too commonly omitted when someone quotes[3] his aphoristic "the only thing we have to fear is fear itself."

In this first decade of the 21st century we seem to have come again to circumstances wherein we once more encounter the paralyzing effect of "unreasoning, unjustified terror." Three generations after FDR's inspiring inaugural address, even the confidence of many Americans that their "great Nation will endure" seems seriously diminished. This is therefore a time for unrelenting inquiry into whatever forces could have contributed to so devastating a loss of nerve. But let us not be misled by the word "terror." Most especially we must not suppose what has had this unnerving effect is simply a "war on terror"—in response to jolting attacks on us by foreign terror*ists.* Some roots of our apprehensive mood are much deeper and far more pervasive; they were in place well before that shocking September day in 2001.

First, though, consider some of the particularly ambient anxieties of a troubled time. It goes without saying that families with young men or women serving overseas in the armed forces must dread each day's casualty reports on the nightly news. Their concern could hardly have been assuaged when the

Secretary of Defense insisted that the country was engaged in "what promises to be a long struggle." Nor did he provide any comfort by declaring it to be a struggle in which "some of the most critical battles may not be in the mountains of Afghanistan or the streets of Iraq" but in newsrooms in metropolitan cities of various countries.[4] These words of his demeaned news reports indicating discrepancies between reporters' perceptions of death and chaotic destruction and the insistently optimistic views asserted by him and his administration colleagues.

But battle deaths and crippling injuries of loved ones in military action were by no means all we dreaded. There was also violent crime, apparently on the rise again in some of our cities.[5] There continued to be a "culture of violence," widely expressed, for example, in appallingly popular video games.[6] Although there was a brief and partial respite from gratuitous violence on American movie screens for a few months after 9/11, the industry soon returned to its habitual and unimaginative ways of using "action" to entice young paying customers into the nation's cinemas.[7]

People dread other calamities as well as violent death. The prospect of serious illness spreading among a population lacking immunity can be terrifying. There was the anthrax scare just after we commenced our so-called "war on terror." The idea of an enemy acquiring biological weapons contributed to a sense of the precariousness of life. Later, that sense became extended beyond concerns about military conflict as news media reported the spread of a deadly type of bird flu from Asia to Europe, Africa, and elsewhere. The causal virus was carried by migratory birds. That virus, upon arrival in another country, began infecting its domestic poultry. Although it was repeatedly noted that very few humans, by exposure to infected birds, had yet contracted the disease, it was also recognized that if the virus were to mutate and become transmissible from human to human a global flu *pandemic* could occur.[8] Reminiscent accounts of the deadly global outbreak of influenza at the end of the First World War appeared and contributed to serious speculations as to the repercussions of such a new pandemic. In various countries, preparations by health care institutions and personnel, and by national and local governments, for coping with pandemic if it happens, were publicized.[9] And in Europe some economic disruption was reported to have already occurred, as 46 countries banned French poultry exports following the discovery of a single infected turkey. France ranked fourth in the world in poultry exports, but the scare cost its poultry industry monthly losses of $40 million.[10]

We live in a time when there are threats, some real, some speculative, not just to our physical health but to other aspects of our being. Some of these we still tend to attribute mainly to bad regimes in other lands, but confidence that "it can't happen here" is not always robust. Although China, for example, has had one of the largest, fastest-growing and most active populations of users of the electronic internet anywhere in the world, ranking second to the United States, the Chinese government has sought to curb the liberative effect of internet access by means of a bureaucracy of censors and a system of technological filters.[11] If Americans are accustomed to nearly absolute freedom to browse the internet, and have had virtually unrestricted freedom to post their own ideas and materials on the web, from time to time they encounter indications that these customary freedoms might not endure. Not only can the system be clogged with "spam," and one's own computer can be violated by "hackers," or disrupted by so-called "viruses" and "worms" unwittingly downloaded, there are ominous signs that internet freedom may be precarious because it can be offensive to "powers that be." Free software called Google Earth, that wed satellite and aerial images to mapping capabilities, was a conspicuous instance of a digitally networked world's escalating openness. It made information once carefully guarded now widely available. But there's the rub. Officials in several countries, including both Russia and India, became alarmed because of their reluctance to have images of vulnerable government buildings or military installations made easily available to anyone.[12] One response was a United States government requirement that only "low resolution" images of certain localities may be put online.

There are other roots of anxiety unrelated to politics, conflict, disease, or power. Some people earn their living by inherently dangerous lines of work. Disasters happen. Sometimes a tragedy can be compounded by misinformation. One instance found most journalists, after having done the misreporting, naturally not blaming themselves, insisting the fault was someone else's.[13] These media people were probably no more self-centered and defensive than people in other occupations. But they had published and broadcast what they were told about 12 miners, trapped underground by an explosion, supposedly having been found alive. The *Washington Post* headlined its story: "12 Found Alive in W.Va. Coal Mine." *Newsday* called it a "Miracle in the Mine." *USA Today*'s banner headline rejoiced: "'Alive!' Miners Beat the Odds." Cable TV networks expressed the same jubilation. According to Rita Cosby of MSNBC, there was a "stunning" development—NBC and the Associated Press said they had "confirmed information" that the 12 trapped miners "are alive."

But only hours later came the corrected announcement, more truly stunning. Eleven of the 12 were *not* alive. After the happy news of the men's survival turned out to be tragically wrong, Associated Press Managing Editor Mike Silverman insisted his organization had reported accurately the information provided by "credible sources—family members and the governor." His statement expressed the plight of the other media as well. They had all been victims of the fact that sources upon which they relied had been misinformed. A message from rescuers in the mine had been either misspoken or was misunderstood when relayed to a command center and then to anxious family members. When they, in turn, told reporters, the reporters forgot their customary reluctance about embracing unchecked information, which can so often in a time of crisis be wrong. The sense of relief was so contagious it had apparently overwhelmed normal journalistic caution.

In the communication chain at the mine there was a kind of "division of labor." Somewhere along that chain, a message got distorted, quite unintentionally. But no particular communicator had to feel personally culpable. Supposedly the reporting of false good news wasn't any individual's fault. Everyone had wanted to believe it. Dispersion of responsibility even made uncertain whose obligation it might be to seek confirmation—and sadly issue the eventual correction.

This episode reminded me of the avid listener to radio "soap opera" programs whom I used to quote in lectures about the way mass media function in our society. Research had found that many regular listeners to those daily 15-minute broadcasts (so abundant on radio in pre-television days) actually took the dramas as a source of advice for living. Some listeners even "referred" friends to a particular program series. One listener even suggested such referral was important to her as a way of avoiding responsibility for any grief that might come to the person thus impersonally "advised"—if actions taken as a result of listening went bad. "If I told her to do something, and something would happen, I would feel guilty. If it happens from the story," she said, "then it's nobody's fault."

Threats to Identity

A clinical psychologist has written about our *culture* of "victimhood" with its trend toward widespread shirking of responsibility. As people grow up they learn evasive attitudes, learning to be nonjudgmental toward their own behavior, blaming others for whatever mishaps they experience, remaining too little concerned about what impacts their own actions may have upon others.[14] How pervasive a pattern is this? Can a society afford for it to be as widespread as it appears to have become? It means people acquire a self-conception which automatically construes adversity as "the other guy's fault." Ego is always a victim, victimized by others. One never acknowledges oneself victim-*izing* others. If we learn to regard ourselves all as victims, then acknowledging any culpability (or even innocent causal connection to unwanted occurrences) is like "blaming the victim." And the phrase "blaming the victim" is usually deemed a pejorative expression indicating someone's unrealistic or mean-spirited attitude.

We have seen an intercultural manifestation of something akin to this, when various self-righteous journalists and thousands of offended Muslims reacted as "victims" to each other's "victimization." Supposedly as a way of ascertaining the extent of "self-censorship," a Danish newspaper editor invited artists to submit cartoons depicting the prophet Mohammed, presumably knowing these would violate an important Islamic taboo, and expecting them to be therefore appropriately restrained. After the Danish newspaper, *Jyllands-Posten* published the submitted cartoons at the end of September 2005, their existence became the provocation for vociferous and ultimately violent expressions of outrage in the Muslim world. Ambassadors to Denmark from some Muslim countries were recalled in protest. In January 2006, the cartoons were reprinted in Norway. At the end of that month the Danish paper offered an apology, but on February 1, papers in France, Germany, Italy and Spain reprinted the cartoons. When *France Soir* did so to demonstrate that "religious dogma" had no place in a secular society, its owner removed the managing editor and expressed "regrets to the Muslim community and all people who were shocked by the publication." Muslim countries reduced their importation of products from Scandinavian countries. Finally, in February and March, scores of people lost their lives as protests turned violent.[15]

Perhaps this tells us what really troubles so many of us today. Rather than fearing real physical dangers, some old and familiar, some new and unfamiliar, now we suffer persistent exposure to threats to our *identity* and our *conception of ourselves*. These, more than threats to our biological lives, appear to be at the root of our continuous unease. What so shocked us about the 9/11 terrorism was not that it put us all in physical danger. It didn't seriously endanger most of us. But the event said something about the nature of our world. So it revealed something about all of us as the world's inhabitants. As resentful victims, humans tend to seek revenge, not always appropriately targeted. This human trait underlay America's national response to 9/11, as devised by our government. By thus expressing our urge to retaliate, an element of our nature we may be finding it hard to accept was revealed. Could our dread of identity loss (or of having our unacknowledged nature *revealed*) become a major obstacle to addressing real problems—or even recognizing and understanding them?

I remember the depth of my melancholy in 1963 the day President John F. Kennedy was assassinated. Neither my colleagues nor I, as we discussed the shock of that November day, felt ourselves in personal danger because those shots had been fired in Dallas. But we agreed we "hadn't thought this was that kind of country" and it hurt to be faced with such wrenching redefinition. Even most of those Americans who opposed JFK politically were as despondent that weekend as those who had supported him as President.

When I was growing up, we just assumed assassination of a president was an ugly aspect of our past that could never happen again. A jokingly sarcastic parental exclamation when we kids misbehaved, even a whispered "And they *shoot* men like Lincoln!" was understood as a reprimand, implying that true justice would require *some* punishment for any behavior less admirable than that of Honest Abe. Those words expressing parental disapproval had always seemed amusing in that former context. But in 1963, finding I was a citizen of a country where assassination of a national leader still could occur turned that remembered phrase utterly sour. Later, it was that sort of identity injury I felt again when an assassin's bullet ended the life of Martin Luther King, Jr.

Five years after Dr. King's death it became absolutely clear that such shootings seemed a personal affront to people like me not just because they produced (as they did) a regrettable change of leadership, but because they seriously wounded our self-conception as an American. Eagerly following the efforts of contenders for the Democratic Party's presidential nomination, during the election campaign of 1968, I was fervently hoping Robert F. Kennedy would not be chosen in the primary elections or at the party convention, even though I liked most of what he said in his campaign speeches, and approved of goals he had worked for as Attorney General. He might have done a fine job as president. The problem for me with his candidacy was that I thought it would be a terrible symbolism, in a world fraught with problems arising from population having already surpassed global carrying capacity, for the White House to be occupied by a president who had so conspicuously exceeded replacement-level reproduction. (*Homo sapiens* is preeminently a symbol-using species, as I intend to make clear in the next chapter.) For Robert Kennedy to become president would, to use a more recently overused phrase, "send the wrong message" to the world. He already had, as I recall, ten children with another on the way. But despite all that, when he was gunned down in Los Angeles, after rousing his supporters with the words "On to Chicago . . ." (for the Democratic Convention), I once again felt grave injury to my identity as a citizen of a supposedly decent society.

When a would-be assassin's bullet in 1972 inflicted crippling injury upon yet another presidential aspirant, George Wallace, I heard the news in a hotel dining room overseas where it was mortifying to overhear another diner speak of the incident as "the American way." With such events it was becoming almost an embarrassment to be identified as an American.

Enervating Factors

This problem of threats to personal identity is not, however, uniquely American. On a planet that is being overused by more than three times as many living human beings as there were when I was a boy, many equipped with vastly more potent technology than existed then, and therefore having much magnified resource appetites and amplified impacts upon the land, the air, and the waters of the Earth, few can truly believe as easily as we used to that we can somehow leave our children and their children a better world than we came into.[16] The mass media inflict upon even the people who don't reside in a crowded metropolis the sense of being hemmed in "by all those others." The sense of loss and deprivation is international—and is hardly limited to the impoverished. People who used to assume progress was perpetual and inevitable at least have occasional quiet doubts now. Some are more able than others to persuade themselves that "things will still work out," despite frictions, conflicts, and severe competition for places in the scheme of things. But it is increasingly difficult in many locales for this planet's human inhabitants to shrug off an oppressive *sense of lost limitlessness.*

If we are living in a time when severe loss of nerve afflicts this nation, and perhaps is an affliction even of Western Civilization at large (and beyond), this calls for careful analysis of all conceivably enervating factors in modern life. I propose to show that the advance of industrial-level occupational specialization—the division of labor which has become ever more ramified through the centuries since the Industrial Revolution—has nurtured tendencies inexorably contributing to that loss of nerve. I will try to explain how division of labor tends to push each of us toward viewing others in *instrumental* terms. Such a degrading influence—violating the Kantian principle that people are always to be viewed as ends, never merely as means—can be a deeper affliction than ordinary fear. Whatever chronically endangers our sense of humanness may be more devastating than threats to our physical safety.

I will try to show how that chronic threat to our sense of worthiness as individual human beings is built into the social order of the advanced technology era. I believe modern societies' reliance upon exchange among specialists does more to loosen than to tighten social bonds. It unleashes forces that continually threaten people's confidence of being personally significant. If a perpetual anxiety about being diminished by some chronic identity-threat does indeed fundamentally flow from living with extreme division of labor, then we are in a potentially insoluble dilemma.

As our hominid ancestors gave rise to *Homo sapiens*, the traits that make us fully human emerged and opened new possibilities for the life of this species. The highly ramified division of labor that has arisen by our time, so much more elaborate than that of a few generations earlier, was perhaps the inevitable consequence of the very attributes that make us human—our incomparable capacity for linguistic interpersonal communication, for fluent manipulation of symbols in systems that range from art to mathematics, for the accumulation and transmission of an ever-expanding cultural heritage, so rich it almost seems to supersede rather than just augment and supplement genetic means of transmitting traits from generation to generation. When we became this human species, we were *destined*, with the advance of time, to produce fantastic technology and elaborate organization, and these would fundamentally alter our participation in ecosystems upon which our lives ultimately depend.[17]

There is no question that extensive and detailed division of labor is a fundamental feature of populous industrial (and so-called Post-industrial) societies, nor is there any doubt that because of this manifest occupational specialization the people in such a society experience a high degree of interdependence.[18] As interdependence deepened there was a reverse side of that coin to which we should have given more thought—seriously diminished self-sufficiency of individuals and local groups. Focusing on the obverse side of the coin, Emile Durkheim, a founder of sociology in France, took an optimistic view in the final decade of the nineteenth century, contending division of labor was a societal blessing, the basis for a new and vital form of social cohesion. Focusing instead, as the twentieth century got under way, on the reverse side—vanishing self-sufficiency, American sociologist E. A. Ross construed industrial-level interdependence as a source of rampant vulnerability.[19] Division of labor led to burgeoning opportunities for people to do grievous harm to one another.

In the chapters ahead I will attempt to resolve the implicit "debate" between Durkheim and Ross. Why is their difference of view important? I worry that what enables societies to grow may be what dooms them to outgrow the planet's life-support systems. But even if we somehow could escape that trap, what has made us human may go on to make us incorrigibly inhumane. It seems to me the "quasi-speciation" within a human population that our culture-producing nature has fostered and made inevitable has the effect of alienating us more and more from one another.[20]

The very attribute of *Homo sapiens* that makes our complex societies possible may thus also make them self-destructive—especially if alienation from our conspecifics (our fellow humans) prods us to continue mistaking vengeance for justice. That is an error that may have been neither so erroneous nor so dysfunctional long ago in our tribal past. In interactions among non-human creatures, violent lashing out in response to being disturbed may serve to diminish recurrence of the other's disturbing action, and such a response may have served (and continued) among early humans. That response today can lead to disaster.

That Simple World No Longer

The evolutionary process—as environmental conditions selected among our hominid ancestors' genetic alternatives and produced the species *Homo sapiens*—is what made it possible for another form of evolutionary process to arise: selection among alternative *cultural* innovations.

In the process of biological evolution, selection pressures are exerted by *existing* (rather than possible future) circumstances. But circumstances change and adaptations become obsolete.

Our species, equipped with culture and the ability to manipulate symbols, has acquired *some* ability to foresee probable future circumstances. So, in principle, *cultural* selection pressures could be future-adaptive, more or less. But because we are cultural creatures we tend to be ethnocentric.[21] We tend to accept familiar ways of behaving as "natural," and to perceive known circumstances as "normal" and therefore as essentially lasting. Our foresight is thus limited. Like non-human species, we continue to seek adaptation to the circumstances we presently confront, not the circumstances posterity will face.

But our myopically adaptive actions are helping to produce the drastically changed conditions in which our descendants must live. The average human capacity for foresight has lagged behind our collective power to change the world.

CHAPTER 2

How We Became Human

"The acquisition of meaningful speech is among the most important learning achievements required of each child. It makes possible the system of significant communication on which virtually the entire structure of society rests."

—R. E. L. Faris 1952. *Social Psychology*. p. 181

"Human hands ... turn out to be wonderful devices for playing the piano; but that's hardly why we acquired them. [O]ur cognitive capacities, epitomized by our linguistic abilities, ... mark us off distinctly from all the millions of other creatures on the planet."

—*Ian Tattersall 1998. Becoming Human, pp. 108, 233*

In 1902, a man with two divinity degrees, who believed in the universality and dependability of both natural law and moral law—each in its respective sphere[22]—became president of Oberlin College. As president for the next quarter century, Henry Churchill King advanced Oberlin's attainment of an outstanding reputation in the field of American tertiary education. This college in a small Ohio town was already internationally known for having pioneered, three generations earlier when newly founded, egalitarian *co*-education of men and women students. In decades prior to emancipation, Oberlin had served as a "station" on the "Underground Railroad" assisting fugitive slaves to escape to Canada. As this educational institution matured it had developed an ambitious curriculum in arts and sciences, and included also an outstanding conservatory of music and a school of theology.

Early twentieth century years were known as the Progressive Era and the predominant theme in King's writings as college president was personal character building.[23] In addition to the extraordinarily diligent pursuit of their studies demanded of students in each of the college's units, President King encouraged a rich array of extracurricular organizations and activities, believing every student should be confronted with an assortment of attractive opportunities far too extensive for any individual to pursue them all. He wanted young men and women to experience the need to make deliberate and thoughtful *choices* in their use of finite time.

Accordingly, when generous support from the federal educational benefit for World War II veterans under the "G.I. Bill" enabled me to be a student at Oberlin in the late 1940s, we undergraduates were kept sufficiently busy that any time gaps for the social life people of college student age are hormonally pressured to desire were precious few. This led to a custom we called "libe dates"—a male student and a female student electing to spend non-class hours at the college library studying side by side or across the table from each other.[24]

On one such occasion I spent a few pleasant moments looking over the top of my book to contemplate the *hands* grasping the book across the table from me. Her hands were slimmer and smoother than mine, and set me to thinking about what a marvelous device the human hand really is. Briefly I pondered the fantastic assortment of manipulations this living appendage at the end of a human arm can accomplish. A hand can point to distant objects. It can wield tools of great diversity, or cutlery at a dinner table. A pair of human hands can be trained to type, or to make music, or to perform surgery, to shuffle cards, to thread a needle, assemble a watch, tie bow knots in shoelaces. And on and on.

Much earlier, when I was barely six years old, one of my hands had been paddled with a teacher's ruler when, at my little desk in the first grade classroom, I inappropriately exercised an acquired skill of my two hands in folding a piece of paper into a flying dart, which I launched toward a friend across the aisle. "We don't throw paper in the classroom," my teacher had scolded. Amid the momentary pain felt as her ruler struck the palm of my reluctantly outstretched hand I had felt resentment toward her inability to recognize the folded paper "airplane" as "in flight," not just "thrown"—in her way of perceiving the event [25]

I don't remember whether or not I noticed during that evening of reading in the Oberlin library the fact that all of this lovely study-partner's fingers were unadorned, but less than a year later it was my privilege to adorn one of them with a diamond ring, and a year after that with another gold band—still in place today 59 years later. On that particular evening, though, just gazing at a pair of human hands set off a train of thought scarcely focused on whose hands they were—a train of thought instead focused on the idea that our hands have much to do with what is involved in *being* human. An idea not altogether new.

Handed Down from Prior Primates

Within a year or so of that ruler-paddling I had heard some adult refer to "the opposable thumb" as one of the distinguishing traits of a human being. But of course I had also known as a boy that other primates have hands in varying degrees similar to ours. It was later that I was told our ability to walk on two legs instead of four had liberated human hands. Precisely when or where I first actually saw a live monkey and marveled at its structural resemblance to myself I cannot say, though I think it may have been in Chicago on a summer day in 1933 when my parents had taken me and my little sister to a "world's fair" (the Depression era's earnestly-optimistic *Century of Progress* exposition, commemorating the centennial of that city). Our parents also took us to a few of Chicago's more permanent attractions, such as the nearby Field Museum of Natural History as well as the Lincoln Park Zoo.

That would have been just after I had completed the second grade, when a teacher I liked much better than the one who wielded rulers to discipline first graders, had read aloud to the class (on special days when we'd all been well behaved) a chapter from a book about *The Early Cave Men*.[26] Those chapters were exciting stories describing the lives of a small tribe of early stone-age humans coping with the challenges of surviving without artificial shelter, with a very minimal assortment of artifacts, eking out a living by hunting and gathering, and eluding dangerous predators. One of the thrills of the Chicago visit was seeing behind glass in the Field Museum a life-size diorama depicting members of a Neanderthal family in front of their cave. The stories our teacher had read to us included an exciting climax in which the early cave dwellers successfully escaped dire danger by killing the menacing "saber-tooth tiger" that had been prowling the region near their cave.[27] So I was especially thrilled to have seen in the Field Museum the fossil skeleton of a real saber tooth, and excitedly reported this to my former teacher the following September. I did not know at the time anything about differences between *Homo neanderthalensis* and *Homo sapiens,* nor was I aware that the saber tooth cat species were already extinct before the people in the storybook were supposed to have lived.

My reason now for relating these reminiscences is to set the stage for some points that need to be considered about human nature and its evolutionary basis. In my high school years in a town in western Michigan, when it was my juvenile ambition to "become an aviator," I used to imagine it would be fun to have been a seagull, for I admired the ability of these birds not only to fly but to glide and to soar, as I watched them use the up-currents of air created when winds coming across the big body of water were pushed upward over the large sand dunes behind the beach on Lake Michigan's eastern shore.[28] Insofar as most of us most of the time are *glad* we happen to be human, one good reason for so rejoicing is that unlike the seagulls, our forelimbs retain a versatility theirs lost in being adapted exclusively for flight. *We* fly by strapping on an airliner, which we unbuckle and "remove" upon arrival at a destination airport. Our hands remain free en route and after arrival to do myriad hand things. This is possible because, unlike the bipedal dinosaur and *Archaeopteryx* ancestors of today's birds, our ancestors were ape-like.

It is because primates have been largely forest dwellers that we have hands. Hands evolved as devices enabling primates to live expertly among the branches of trees. Fingers that could hook over branches enabled swift and skilled climbing, especially when aided by a somewhat opposable thumb for grasping. The fact that the *feet* of many primate species are shaped so differently from ours attests to this arboreal way of life, and caused apes and monkeys to be referred to in the past as quadrumana (creatures with four hands). Unlike birds, the primates could not fly, but also unlike birds they could grasp insects, fruits, nuts, and many other objects, edible or otherwise useful or interesting. Grasping small objects with hands has advantages over the grasping a bird can do (with its bill—a mouth appendage—differing in shape according to the way of life of a given bird species). Even mammals, ones that are not primates, and thus lack hands, cannot as easily carry objects (including their own offspring) from place to place. Even as a small child I had known that a mother cat, for instance, has to transport her kittens, one at a time, by grasping in her mouth the loose skin on the back of the kitten's neck.

One of the playground devices next to my grade school was a ladder-like structure horizontally mounted atop some upright columns, suspending it perhaps six feet above the playground surface. This device was called "monkey bars," and the recreational use of it involved climbing up to grab one of the "rungs," then reaching forward to the next one, and swinging along from rung to rung with feet off the ground. When apes do this among the branches of trees we call it *brachiating*. At that age I was never an accomplished brachiator but I admired some of the older boys who could swing along swiftly all the way from one end of this playground structure to the other.

Perhaps if I had been growing up in a Southeast Asian community near the habitat of the little siamang apes, or other gibbon species, with their truly impressive brachiating ability, they, rather than seagulls, would have been my preferred alter-species identity. But all things considered, I'd rather not be a gibbon; their hands are fully adapted to that activity, with much more elongated fingers, and lesser thumbs. Hands like theirs would have a harder time using a pen or pencil or a knife and fork.

However, unlike the 19th century's Anglican Bishop Samuel Wilberforce, who was so incensed by Charles Darwin's insights about natural selection, I have no resentful urge to deny ape-like remote ancestry.[29] I *rejoice* that it bestowed upon me these hands—modified through the ages by intermediate generations bereft of the former forest habitat, who invented and used tools as they adapted to life on the African savanna, resulting in hands changed to an extent where my fingers feel well suited to touching my computer keyboard, using toothbrush or comb, turning the pages of a book, lacing up hiking boots, or pressing a camera's shutter-release.

I cherish also my earlier tree-dwelling primate ancestors' additional legacy of forward facing eyes, for with this arrangement the range-finder convergence by my two eyes enables me to see the world in all its three dimensional form. Together with the marvelous convenience of hands, and apelike depth perception, I appreciate inheriting the *color* vision also achieved by these simian forebears. These attributes are all wonderfully useful and often thrilling.

With arms somewhat longer than hind legs, the quadrupedal movements of African apes take the special form called "knuckle walking." Their hand-like rear feet go more or less flat on the ground, but the hands on their longer forelimbs reach down with fingers curled enough so that it is their knuckles that touch the ground and bear the anterior portion of the owner's weight. Accordingly, the wrists of gorillas and chimpanzees differ from ours and their fingers are shaped somewhat differently from ours, and would not likely be able to acquire quite such delicate manipulatory skills as human hands have acquired—to thread a needle, assemble a Swiss watch, or dissect a specimen in a biology lab. More significantly, pre-human primates could not even know what these actions meant.

Having Hands Enhanced

The African great apes are able to accomplish substantial spurts of two legged walking which does indicate that the *common ancestor* of them and us, between 10 and 5 million years ago, could likewise walk upright part time. That was a preadaptation which enabled the lineage leading to us, after it split off from the African apes' lineage, to evolve *full-time* bipedalism. It turns out that the attainment of full-time bipedalism occurred quite early in the process of becoming human.[30] It preceded the cranial enlargement that signified in the fossil record the emergence of another distinguishing feature of humans, our brain. Why this was the sequence will become clear in a moment. First, it is important to recognize that not all instances of bipedalism produced the same result, and the reasons for different outcomes are clear.

Although for members of the ancestral ape populations full-time bipedalism was an innovation, it was not at all new among nature's creatures generally. Birds are bipedal, and before birds, their reptilian ancestors were bipedal. Many species of dinosaurs walked and ran on just their hind legs, and had smaller arms with smaller "hands" (often with only three sharp-clawed "fingers"), more suited for tearing flesh or vegetation than carefully "handling" any small object (much less any intricate "device"). They could not have engaged in social "grooming" as monkeys and apes do, picking insect parasites off of one another.

The forelimbs of reptiles, birds, and of most mammal species other than primates, were adapted for other functions than grasping and manipulating things. At the extreme, for example, the flippers of aquatic mammals, and the wings of birds were adapted to the specific tasks of swimming or flying. These adaptations rendered them *unavailable* for development into manipulatory devices like our hands.

When our remote ancestors stood up to walk, this novel but advantageous form of locomotion had the incidental effect of leaving their hands more free than those of their (quadrupedal) ancestors for doing other things.

Full-time bipedalism thus enabled our earliest hominid ancestors, who already had ape-like hands, to begin a process of evolving hands perceptibly different from those of apes. The hands of succeeding generations of our ancestors would become ever more like ours, as subsequent ways of *using* such hands evoked additional selection pressure for development of further human traits—specifically the larger brain, and in time, the very essential ability to speak. When our species ultimately acquired the capability for elaborate vocal communication, the potential for ingenuity and innovation became enormous.

We *had to* have descended from apes to have our versatile hands. But further, the particular types of primates we call hominids already lived in groups, not as totally isolated individuals who came together only occasionally for mating. Over many generations, as the early hominids' descendants *perfected* the habit of upright walking, and as individuals who happened to be a little bit more favorably constructed for walking that way were enabled to thrive somewhat better than others whose legs were less well adapted for bipedal walking, they would thus leave, on average, more surviving descendants.[31] With the advent of full-time bipedalism, the hands of hominids became more available for carrying things.[32] Whatever foodstuffs they obtained as they foraged out in the open could be carried back to a more sheltered place to be consumed—either by themselves or sometimes to be shared with associates.

Food sharing became a context in which new influences upon one's life chances would arise. The uses these protohumans made of their hands would have been further elaborated, little by little, generation after generation—especially by those who happened to be born with slightly more dexterous fingers, and who somehow had finer control of the movements of hands and fingers—for making and using tools. They in turn would have bestowed their genes on a slightly increased fraction of the descendant population. Structural changes would have happened according to the process so aptly labeled by Charles Darwin as *natural selection*. We call the cumulative change evolution, but Darwin used the descriptive phrase: descent with modification.

As in any population of any species, there was certainly appreciable variation between individuals in size, strength, external shape. But there was just as certainly some inter-individual variation in the nervous systems, in the brains, of these protohumans. The circumstances of life for these creatures would have exerted *selection pressures*. Factors of selectivity enabling some individuals to achieve more reproductive success than others would, in time, because the advantageous traits were often heritable, cumulatively change the nature of the species population.

One Change Leads to Another

Freer and improved hands could do more things, including more delicate manipulations. However, the beneficial effects of becoming bipedal creatures with useful hands no longer committed to knuckle-walking meant that our ancestors needed changed neural mechanisms for most usefully controlling the movements of their hands and fingers. Because the myriad parts of an organism are connected and affect each other, improvement of hands, and ever-more varied use of hands, would reshape population averages *in other parts* as well as hands and legs and arms. Changed behaviors and changed structures went "hand in hand," so to speak. Natural selection would eventually endow creatures possessing advantageous hands with *modified nervous systems* that could put hands to optimal use. Over time, these modifications would include increments of brain tissue, and the change would be discoverable in our time by measurement of the cranial capacity of fossils, carefully dated by various methods (including initially, but now no longer limited to, estimates of the age of the strata in which the fossilized remains were found).

Living in small collectivities the earliest hominids also needed means of communicating with each other, not just about dangers and hazards but also about opportunities and intentions. To some extent, hand gestures would certainly have been put to communicative use.[33] Like their forest-dwelling ancestors, these protohumans were equipped with lungs, chest muscles, and a vocal tract that enabled them to produce various sounds that also already had limited communicative use—cries of alarm, pain, anguish, rage, or calm satisfaction. Today's human vocal tract differs from the vocal tracts of apes in ways that enable production of a more elaborate repertoire of sounds. Voluntary control of the small muscles in lips, tongue, and larynx became important to enable increasing communicative use of sound, so *selection pressures* associated with the processes of group living eventually brought about pertinent additions to brain circuitry. That is to say, brains able to make better use of hands, and brains able to begin elaborating the communicative use of vocalization, would have given advantages to those who happened to possess such brains.

The brains of today's humans are equipped to enable language to be learned and used. This came about by natural selection, operating through many generations that lived between five million years ago and a few tens of thousands of years ago. Normal human beings today, wherever in the world they happen to be born, learn to use one language or another.

How do we do it? First of all, infants almost from birth *play* with their various muscles. Movements of arms, legs, fingers that are initially random become deliberate as the child grows. The human child also *plays* with its vocal tract. As infants we babble. Presumably, some of our hominid ancestors, as infants, did too. Living in groups made it "safe" to babble, as the group provided substantial protection from whatever predators might have been drawn to a helpless infant by its vocalizing. Babbling today seems to be an essential precursor of language learning by children, so it doubtless helped earlier generations to acquire speech.[34] The baby's earliest vocalizations are limited to a very few of the easiest-to-make sounds, but its playing with its vocal apparatus gradually expands the assortment of sounds it can produce. Soon, babbling becomes less random as the playing focuses increasingly upon emulating whatever language sounds the child may hear in his or her surroundings—in the speech of the parents, other adults, and older siblings. Babbling serves as the warm-up to the baby's own beginning speech efforts. It is a prelude to the infinite sentences to come.

Children who happen to be deaf (from birth, or from early illness), although they show some signs of vocalization, do not make sounds that function like the babbling of infants with normal hearing. Without hearing they cannot focus on emulating sounds produced by others. Systematic syllable sequences do not emerge for deaf children as they do for hearing children. However, deaf children *do* babble. But they *babble with their hands* rather than vocally. If they are exposed to sign language, their spontaneous hand gestures will go beyond mere shaking and waving, etc. Specific hand shapes, locations, and movements in the sign language adults around deaf babies may be using, will tend to be reflected in those babies' own movements.

It is a striking fact that manual babbling by deaf babies emerges at the same age as does hearing babies' vocal babbling. Like the latter it initially lacks meaning. It is rhythmic and repetitive, and develops into combined patterns. Then, at the age when vocal babbling normally begins turning into speech, in the right social context such manual babbling will begin turning into meaningful manual signs. All human children have the combinatorial capacity to make use of a finite set of symbols—be they consonants and vowels, or hand shapes—to generate abundant "sentences."

The parallel facts about oral and manual babbling and the early stages of learning to talk or sign seem to indicate a connection (or at least a close similarity) between the neural apparatus needed to control our hands and that required to control our vocal tract.[35] This impression is strengthened by the fact that even as adult speakers we persist in making hand gestures (often unwittingly in ordinary conversation, and even on the telephone when unseen by the listener) as we emit words vocally.

Vocal Tract Gesturing

The possible connection's plausibility is further reinforced if we spend a few moments exploring our own vocal musculature in action.[36] If we consciously try, we can actually *feel* how we speak—by saying *aloud* such words as fate, date, rate, mate—making an effort to notice how the tongue and lips behave in pronouncing each word, and even how the shapes of mouth and cheeks change. Then say pate, and bait, and recognize the difference in the timing of "turning on" the vocal cords—the voice comes on a split second *after* pronouncing the *p* and it comes on just *before* pronouncing the *b*. Likewise say (and think about the process of saying) kill—and then gill. What did you feel happening differently in the back of your mouth? Now think about what is different in pronouncing the *th* in moth and in mother. Finally, think about the difference in pronouncing fuss and thus, or missile, whistle and thistle.[37] And notice what your facial muscles do when you say hound, and round.

When humans talk, we make amazingly adept use of surprisingly complex biological apparatus. Clearly, we have the ability to *control* (usually with marvelous precision) the various parts of this vocal tract structure.[38]

So the operation of natural selection across thousands of generations reshaped hands, (together with arms and legs), and as a consequence of their changed use, heads (and both the vocal and mental contents thereof) came to be reshaped as well. We differ substantially from all living apes. Our larger brain is not the *only* distinction that makes us human, and its enlargement resulted from earlier attainment of other human traits.

Toward Enhanced Knowledge

For students of my generation at Oberlin, the "open stacks" policy that gave them freedom to browse throughout the college library was a truly wonderful opportunity for the use and on-going development of their advanced primate heads. Assigned searches or unplanned explorations could broaden intellectual horizons. One of the books I happened to read that made a lasting impression was by a Cornell University historian, Carl L. Becker. Originally published in 1922, its title was *The Declaration of Independence*, but its subtitle is indicative of how it altered my way of thinking—*A Study in the History of Political Ideas.*[39]

Its fourth chapter, entitled "Drafting the Declaration," was particularly memorable. That chapter included the text of a "Rough Draft of the Declaration" as submitted by Thomas Jefferson to his drafting-committee colleague, Benjamin Franklin. Differences between that rough draft and the familiar-to-most-Americans finished declaration were easily spotted. Changes suggested by Franklin and by another member of the drafting committee, John Adams, were shown as interlineations. There followed another revised Rough Draft—being the "fair copy" that was submitted to Congress as the committee's report—showing interlined corrections and additions Jefferson himself made to the Adams copy. For comparison, the chapter in Becker's book included also the text of the Declaration as it reads in "the parchment copy" officially proclaiming separation of the thirteen colonies from the mother country.

Reading Becker was effective in disabusing me of the simplistic notion that great and inspiring ideas appear from mere genius, born fully mature without prior gestation. Considering the process and the steps by which ideas actually did develop may be essential for understanding what they are and what may come of them. And as I learned many years later, this is equally true for understanding organisms, species and the ecosystems comprising them. Studying earlier versions of each is informative about present instances. We need to know how they came to be as they are. This knowledge helps us understand why they function as they do—and may yield insights regarding future consequences of their existence and on-going actions.

Finding out about steps in the development that preceded the finished form (of other documents and other entities) turns out to be both enlightening and sometimes fascinating. Fortunately, equivalent studies tracing origins of great ideas appear from time to time in fields other than American history. One better understands, for example, the theory of *evolution by natural selection* the more one learns about earlier ideas that led up to it. It also clarifies matters if one learns about the earliest incipient formulations of the principle itself. And it helps to know about the masses of pertinent evidence accumulated *since* the time of Darwin.[40]

When scientists in the 19th century first began to pursue implications of our descent from ape-like ancestors, the only kind of relevant evidence available was expected to be the fossil record, then still extremely sketchy. Searches would have to be conducted for fortuitously preserved (and then yet to be discovered) skeletal remains of whatever now-extinct species might constitute links of descent from a common ancestor of ourselves and chimps down to the present time when these apes and we are altogether separate species. Those who sought fossils to study were exploring hypotheses that were inevitably shaped by some assumptions derived from traditional pre-Darwinian thought ways. Because the most important distinguishing characteristic of humans seemed so obviously to be the much larger brain of *Homo sapiens* compared to any living ape's brain, early fossil hunters who hoped to unearth specimens left by early hominids expected that these might resemble ape skeletal parts more than they resembled modern human bones, except that they would "of course" have a cranium conspicuously larger than any contemporary ape's. It was that expectation which enabled the Piltdown hoax to occur.

So-called Piltdown Man was announced in 1912, represented by a cranium and part of a jaw found in an English gravel pit. The specimen seemed to represent a creature with a very human brain cage and a very apelike face (as indicated by its jaw). Scientific inquiry was confused by this "specimen" for years. The purported hominid, named *Eoanthropus dawsoni*, to honor the finder, Charles Dawson, seemed to confirm preconceptions (and by having "turned up" in England served British ethnocentric pride). But it simply could not be fitted into patterns induced from other finds elsewhere.

It was not until 1953 that the Piltdown "find" was exposed as fraudulent. And later still the "whodunit" was solved.[41] The Piltdown remains had consisted of recent bones made to look ancient and clandestinely placed where they would be "found." A human cranium only a few hundred years old had been planted with a misleadingly modified jawbone from an orangutan. Taken together, they gave the expected (but terribly misleading) impression that the large brain had preceded all other prehistoric human evolutionary developments. When Piltdown Man was proven to have been a hoax (and was expunged from the fossil record relied upon to understand hominid evolution), the picture of human antecedents became much less confusing. The real sequence of changes that had to have occurred between *Australopithecus* and *Homo sapiens* began coming into sharp focus. Upright walking came first, accompanied and followed by increased use of hands for making and using tools, then rudimentary communication was required by group living. All of these implied selection pressure favoring new neural circuits. Brain-size increases followed, and eventually speech.

What We Have Become

Although, as the linguist Charles Yang points out, *every* living species "is a testament to the power of evolution," humanity's achievement of the power to communicate via language is special. The ability of our preliterate human ancestors to cooperate in hunting, to make and use tools, had conspicuously simpler precursors in other species. But, to the best of our knowledge, human language far excels the varied systems of limited communication employed by other animals.[42]

It is reasonable to consider those animal communication systems as having foreshadowed human language, especially when we credit apes in the wild with a substantial degree of gestural and vocal communication and add to this what has been learned in laboratory experiments with chimpanzees, bonobos, and gorillas, about their mental abilities and activities in humanly contrived situations. But it is also not only reasonable but imperative to recognize that our human capacity for language is so enormously further advanced quantitatively as to constitute a difference in kind. This was true even before the invention of writing. Eventually printing and the mass dissemination of knowledge and ideas in books, not to mention electronic recording and replication of printed materials, have enormously extended our species' difference from all others. It is little wonder, then, that people think of themselves as apart from "the animals," and "above" nature—whether or not they invoke sacred scriptural justification for their anthropocentric attitude and assumptions. Language begets hubris. Does it also enable *recognition* of our hubris—and retreat from its perils?

Not being able to know in depth what goes on in the minds of other animals, we tend to suppose the concept of *mind* is scarcely applicable to nonhumans. But when a dog chases a cat and the cat escapes up a tree, at least the fond pet owners may impute thoughts to their two animals—as if the dog had said to itself, "I spy a cat, which I will enthusiastically chase," and the cat said to itself, "Here comes a dog, much bigger than I am, hence dangerous, but there's a tree which I can climb and it can't."

But dogs and cats *don't* verbalize. Does that stop them from *defining* in some way the objects they encounter? Do dogs—from some instinctual urge bestowed by their wolf ancestry—somehow perceive a cat, in effect, as something to be chased, rather than as a puppy (of its own species) to be played with or nurtured? Does the cat *define* the tree (in some nonverbal way) as a climbable refuge? Would it dash to the tree and climb it to escape the dog if it perceived the tree merely as a source of shade?

For humans, with language, we can express in words the ways we perceive things and events and possibilities. Accordingly, sociologists and social psychologists have long insisted that we respond to situations according to our "definitions" of them.[43] We define situations with words. (If dogs and cats and apes define situations, they must do so by mental images without using words, and this has to limit them seriously in comparison with humans.) With words, we humans can have many more (and more intricately detailed) definitions of things and opportunities, and of experiences, even of other people.

From Language to Culture

Language attainment (or the ability to make and use symbols, in general) was the ultimate breakthrough in human evolution. Language use can malfunction, even though it was the great enabler of social organization, cultural elaboration and innovation. It made possible the development of religious doctrine, philosophical thought, scientific conceptualization and research.

Before there was language, cultural innovation was exceedingly slow and limited. Without language there would be only simple division of labor, little more than task-specialization by age and sex. There could be no rich diversity of artifacts, no mass production, no rising standard of living (as economically defined).

As humans increased in number, and spread themselves around the world, with many local populations becoming effectively isolated from other populations elsewhere, an enormous differentiation of languages naturally developed. Just as geographic isolation has played a fundamental role in speciation, geographic dispersion of *Homo sapiens* had to result in the proliferation and divergence of languages. Language differences became barriers to interaction, and could function much like geographic isolation in interrupting gene flows (i.e., obstructing interbreeding of separate groups). All this enabled us to define ourselves as "peoples" (plural)[44] rather than seeing ourselves as a single species, all members of which, wherever they may reside and whatever culture and social system may be their matrix, share common interests such as the availability of reliable life-sustaining surroundings, and possibly concerns for quality of life, assurance of liberty, and opportunities to pursue happiness.

Homo sapiens is unquestionably a cultural species, with behavior shaped by learning processes that are intrinsic to societal living, by one or another social heritage as well as by genetic heritage. Different populations speaking different languages *and* located in different places on this planet were bound to evolve different cultures. Different cultures will bestow upon the people living by them different habits of mapping experience with words. Whatever commonalities may be found in all cultures—as they must all reflect the fundamental abilities and limitations of the human species—differences between cultures have become a basic fact of human life in our time. Some of the differences may appear to have become such unresolvable incompatibilities that, like the stereotypic sheriff in a Hollywood western movie whose territory was invaded by some notorious outlaw, people of one culture may feel toward another that this planet "ain't big enough for both of us."

Territoriality characterizes the lives of various non-human species. Lethal conflicts sometimes occur even between one band of our cousins, the chimpanzees, and another band of those cousins. But the total chimp population is far, far less than the total number of human Earthlings. Any threat of extinction facing chimpanzees due to habitat damage is not of *their* doing. It is the progress of *Homo sapiens* that may terminate the existence of *Pan troglodytes*, by depriving those cousins of the habitat to which they are adapted.

Language is for humans much of the time a wonderful advantage. But the advantage can be fraught with disadvantageous side-effects. Human language use can malfunction. Words can be ill-chosen. Words can *mis*-define situations. Our words can mislead others—and ourselves. And we may not always recognize when we are being misled. The question lurking now in all our accomplishments is this: Has the acquisition of language and the general ability to make and manipulate symbols produced a "human nature" that remains potentially sustainable, or could it prove ultimately self-destructive? Will the language ability that made us, also destroy us?

Toward answering that question, three aspects of the evolutionary process are worth emphasizing more than is usually done.

(1) Organisms tend to exploit niches to which they are adapted whenever and wherever such niches become available to them. In short, life is *opportunistic*.

(2) Natural selection happens in response to the *present* availability of niches in a given environment, without regard for their possible future availability.

(3) *Organic prescience is typically very limited*—organisms typically do not foresee future environmental conditions. Living things adapt to what *is*, not to unseen future circumstances. If those are drastically different, the obsolescence of today's successful adaptations may then prove fatal.

From these three principles an important pattern follows. Opportunistic occupancy of niches can lead to burgeoning populations of the species adapted to them, which sometimes results in surpassing the given environment's *carrying capacity* for such organisms. In other words, organisms, by using the available environment change it—reducing formerly abundant niches, generating different niches. Use of a given environment may in time thus make it less fit to support the organisms using it. They and their progeny may be unable to cope with the conditions of the changed environment.

The words people exchange may or may not adequately convey recognition of such changing circumstances. So the question really becomes: are language-using humans enough different from other species to escape this process of self-destruction by habitat destruction?

CHAPTER 3

Why We Deceive and Can Be Deceived

"Great though may be the leap we have made away from the rest of the living world in the acquisition of symbolic thought, we have not entirely emancipated ourselves from the brain structures that governed the behavior of some very remote ancestors indeed. And it is precisely this interaction of the ancient with the new that makes us not only unique in many admirable ways, but also uniquely dangerous—as much to ourselves as to the rest of the living world."

—Ian Tattersall 1998. *Becoming Human.* p. 234

When I was an undergraduate student, the World War II G.I.Bill made it feasible to attend classes right through the summers, rather than having to seek a summer job. Oberlin did not have a summer session, and at that time there were still a few regions of the country with which I was not yet familiar, so prior to my junior year I chose to enroll for a summer term in New England—as it happened, at the University of New Hampshire, located in an even smaller village than the Ohio town of Oberlin. I was able to take a pair of courses there and have the credits transferred back to Oberlin. One course was in the English Department, taught by a professor whose name I do not remember, but whose attitude about language and its creative use I do remember. To him, language was humanity's most vital tool. Words, he insisted, should always be used with great respect. We were taught to avoid using words in cute, clever, or tricky ways that might damage their ability to convey meaning. At the very least we must be careful with any metaphors we choose to employ, so as not to bend and break the language tool.

Humans as Word-Cartographers

One of the books to which his course introduced me was *Language in Action* by S. I. Hayakawa, then a professor at Illinois Institute of Technology in Chicago, later at San Francisco State, before becoming a U.S. Senator from California. Hayakawa's book introduced me to some concepts under the heading of "General Semantics," including the idea that our use of language in defining the situations we encounter in life is analogous to the use of maps when we travel, or when we simply want to understand the locations of places and the characteristics of regions. The phrase "word-maps" began to serve as a vivid expression of the relation between language and experience. We do "map" the world we experience when we describe it in words. Our verbalizations should have a clear correspondence with their referents, Hayakawa was telling his readers. The relationship between words and things or events should be like the relation between a map and the territory it represents.

But maps can be flawed. Our definitions of situations can sometimes be erroneous, or inadequate. Language is for humans much of the time a wonderful advantage. But the advantage can be fraught with disadvantageous side-effects. Words can be ill-chosen. Words can *mis*-define situations. The word-*maps* by which we find our way around in life can sometimes be misleading, even those we construct ourselves. And we may not always recognize when we are being misled.

The following semester, back at Oberlin, my roommate introduced me to another, more elaborate, treatise on "General Semantics" by Alfred Korzybski, *Science and Sanity*, which established the word-map concept more deeply in my habits of thought. I also began to think about different "levels of abstraction." Such words as "chimpanzee" and "gorilla" can represent a particular level of abstraction, and to speak of "apes" (a more general category than its member species) would be a step up the "abstraction ladder." "Primates" would be another upward step. "Mammals" would be on a higher rung than that, and "Animals" would be more abstract still. On up from there we might come to "organisms," etc. Each more *abstract* word stands for a more inclusive array of entities but says less about their particular characteristics. One of the concerns repeatedly expressed by Korzybski—who insisted words can't say everything about anything—was that too much human communication is conducted at too high a level of abstraction, and thus may not be saying enough. Conversations too high on the abstraction ladder omit some specificity that could have relevance. He urged readers to be continually striving to work their way *down* the abstraction ladder, and to remember that *the map is not the territory*—it only represents it, and may sometimes *mis*represent reality.

Word-maps seldom or never tell all there is to know about real situations. Usually this is understood and may not be a problem. But we can become too wordy, trying to convey (or capture in our own thought processes) more information than we can effectively use. Too much can become confusing. So, in fact, it is necessary to keep word-maps uncluttered with superfluous information, just as we avoid confusion when relying on road maps or navigation charts by insisting they not be cluttered with excessive detail.

But sometimes the omitted details turn out to have been more important than their omission from the word-map seemed to imply. That overheard comment in a foreign hotel dining room, characterizing the attempted assassination of a contender for the U.S. presidency as "the American way," was hurtful to my identity because it put the incident too high on the abstraction ladder—omitting from consideration many, many other components of American political processes, some honorable, some indifferent, some despicable.

With whatever language habits and other habits of symbol-using we happen to develop, we may sometimes subject ourselves to self-deception. Recent severe rainstorms in the region where I live have resulted in serious erosion. People living in some hilltop homes, who built where they did because of a "definition" of that situation as prime residential property, attractively scenic and socially desirable, have suffered serious loss when the land beneath their dwelling was abruptly changed, as the hillside washed away. In some instances, residents at the bottom of the hill have also suffered devastation, as the eroding hillside crushed their home or buried some portion of it. Their "definition" of that bottom-of-the-hill situation had been that it was "affordable." Their definition perhaps ought to have been that it was *endangered* by the hillside above it.

I will be referring to word-maps and the abstraction ladder concept again in a later chapter. For now, though, it suffices to say that word-maps are an important component of the way human beings relate to others and to the world. We "navigate" through life with word-maps—and with symbols of many kinds. As a sociologist I have probably used the expression definition of the situation more often than the term word map—but they are roughly equivalent concepts and both convey a sense of the importance of language as a fundamental capability of our species, an essential tool for behaving as human beings.

The Symbol-user Species

Homo sapiens has abilities to use (and be influenced by) other kinds of symbols besides words and language. Mathematics represents our ability to use a special category of symbols—numbers and the mathematical "operators"—to add and subtract, multiply and divide, to develop logical proofs of geometric principles, to understand and manipulate equations, and on and on. A rich "vocabulary" of mathematical symbols and a mathematical "grammar" enables us to go way beyond counting fingers and toes, even if that ancient activity may have been the basis for our decimal system. Progress in mathematics has nurtured endless achievements in various sciences, making plausible the focus on brain-power implied by the name Linneaus gave our species—*Homo sapiens.*

Any use of symbols depends on our species having evolved the ability to let one thing "stand for" another thing. The sound "dog" is understood by English-speakers to *stand for* a canine animal—as if when we hear that sound it arouses in us a mental image of such a creature. But we have long since learned also to use the letters d, o, and g to *stand for* the sound of the spoken word "dog." By reading, we allow our mental images to be stimulated by printed alphabetic symbols on paper (or stone, or a computer screen) rather than by sounds. Or, the letters can, in turn, be represented by dots and dashes in telegraphic code, or by even more arcane electromagnetic impulses inside a computer (or in communication between computers). Or, letters can be represented by patterns of raised dots to be read by sensitive (ape-derived) fingertips of people without eyesight. So word-maps can exist at several levels of abstraction.

Moreover, the symbolizing ability characteristic of our species operates in yet another realm. Long ago, people discovered the advantage of having some particular commodity *stand for* any of the various items they might want to exchange with one another. The particular commodity they used as "currency"—to represent value in general—could have been small pouches of salt, ornamental seashells of a certain type, or beads fashioned from animal teeth or pieces of bone. Later, specified quantities of a "precious metal" (e.g., gold) tended to become the standard medium of exchange, and eventually people learned to form it, and other metals, into coins. The size and design of the coin could denote the quantity of value it was supposed to represent, and the circulation of such monetary objects facilitated the exchange process without which division of labor could not have advanced very much, and human life would have remained much less diversified than it has become in our time.

But in the realm of money, as in the realm of words and mathematics, one could ascend the abstraction ladder and let something else *stand for* those coins. An accumulation of metal coins might become heavy and inconvenient to carry around. Pieces of paper suitably marked (first as mere receipts, later in more formalized fashion as printed banknotes, could be used as money in lieu of the coinage those pieces of paper represented. Other pieces of paper were devised to serve as "orders to pay" either notes or coins, and came to be called "checks" (or "cheques"). Now, of course, just as we represent words with electromagnetic impulses, we also do this with money. We speak of "plastic money" when we refer to credit cards, which more or less take the place of (or reduce the daily need for) writing checks. We even by-pass the plastic card when we type in the card number as we order something "over the internet." And those rectangular pieces of plastic now commonly have a magnetic stripe along one edge that can activate an electronic reading device, to convey the "order to pay" to some remote location.

When one stops to think about it, it should be obvious that these various levels of abstraction of these several worlds of symbols are truly essential to human life as we know it today. Further, just as we can conjure fictions with words, creating fairy tales, myths, and legends, so the way we manipulate the monetary symbols (on their various abstraction levels) can "conjure fictions"—some eminently useful, some fraught with dangers to ourselves or others or to our common future. When a drop of nine percent in the overall value on China's stock exchange (on a day in March, 2007) triggered selloffs in other nations' markets around the world, people were jolted into awareness of the threatening side of reliance on high-level abstraction in the monetary world. But, I was told, that nine percent drop had closely followed a one-day rise of *twelve* percent on China's stock market—and the astute financial adviser who mentioned this to me went on to ask rhetorically: What could possibly ever justify a 12 percent increase in value in a single day?

When *Homo sapiens* emerged as a large-brained, symbol-wielding species, so that new kinds of within-species differentiation (by culture) became possible, the seemingly always advantageous division of labor between proliferating occupational specialties was a significant result. But advantages can turn out to have disadvantages. In our time we must at last begin considering the disadvantages implicit in our prime human advantage.

From Aesop's 6th century B.C. fable, "The Wolf in Sheep's Clothing," children have long been exposed to an allegorical suggestion that there are circumstances in which deception, even though risky, can be useful (for manipulating others' behavior).

Why Do We Deceive?

When deception is considered in an evolutionary perspective we can recognize that it is occurring whenever a creature of any kind induces in another creature a false or misleading "definition of the situation." That venerable sociological concept—definition of the situation—has wider application than often realized.

In college sociology courses, the "definition of the situation" concept has been presented as a characteristically *human* response process. According to W. I. Thomas at the University of Chicago, for higher animals, and *especially* for humans, "Preliminary to any self-determined act of behavior there is always a stage of examination and deliberation which we may call *the definition of the situation.*" Not only specific acts depend on definitions of the situation, said Thomas: the personality of the individual and a whole life policy develop from a series of such definitions.[45] At birth the human child always enters into a group of people who were born earlier and have been shaped by various experiences. As life goes along among the group's members the various recurring types of situation they are likely to encounter yield shared definitions. Rules of conduct corresponding to those group definitions develop and are handed down.

Sociologist Lewis Coser[46] urged his readers to ponder carefully what he regarded as "the most pregnant sentence" ever written by W. I. Thomas (in 1928): 'If men define situations as real, they are real in their consequences.' Robert Merton called this "the Thomas theorem."[47] People respond, it was telling us, not just to the objective features of a situation, but to the meaning that situation has for them, often culturally prescribed. Their behavior is shaped by the situation's meaning *as they perceive it.*[48] But that is no uniquely human pattern. What may be uniquely human is a tendency for obsolete definitions to persist, prolonging habits and customs after they have become maladaptive.

When and why can deception be expected to occur? Is it always harmful? Deception is not merely a character flaw in criminals—or in some politicians. It is not even uniquely human. Animals deceive each other and even plants can be adaptively deceptive in some fascinating ways.[49]

What is the significance of the fact that there are many non-verbal ways of telling lies, and many non-human provocations for doing so?

Consider a story that seemed funny when I was a boy. It supposedly taught the perils of being naïve. A farmer responded to a nighttime disturbance in his hen house. Shotgun in hand he called through the doorway, "Who's there?" Lurking in the darkness, the would-be chicken thief replied, "Nobody here but us chickens." "O.K.," said the sleepy farmer and went back to bed—at least so said the story's laugh line.

Neither the ability to deceive nor to be deceived is uniquely human, nor is either necessarily something done consciously. For plants as well as animals the ability to play the game of "Nobody here but us chickens" may be essential for survival. Either for a prey animal to escape its predators, or for a predator to capture prey, having the ability somehow to create the impression it isn't there can be vitally important.

Differential Association, Opportunities, and Lies

Suppose the society into which one is born tries to instill a rule that lying is bad and must not be done. Then suppose one tries to conform to expectations of honesty but one is also pressed to follow whatever other expectations are peculiar to one's particular group within the larger society. The more complex a culture has become, the greater the number of normative groups likely to affect a person, so the greater the likelihood that the expectations of these groups will differ.[50]

Were we to insist on trying to explain human deceptive behavior within a strictly conventional sociological framework, we might paraphrase the venerable explanation of delinquent and criminal behavior known as *differential association theory*.[51] We would then say something like this: Exposure to an excess of standards favorable to lying and relative underexposure to standards opposing deception seems likely to enhance the probability of behaving deceptively. Individuals would tend to resort to lying as a result of contacts with deceptive behavior patterns and isolation from patterns of honesty. People who, by reason of their residence, employment, etc., have been relatively isolated from the culture of scrupulously honest groups or have had relatively frequent contact with a culture of deceit would have greater than average likelihood of lying, having lacked the experiences, feelings, ideas, and attitudes from which they could construct a life organization that would comply with public expectations of honesty.

But this would only explain *which* people can be expected to deceive, if anyone does. Even for them, however, for deception rather than honesty to be, or seem to be, advantageous, *opportunities* to deceive must be perceived.[52] As societies change there are changing patterns of opportunity. Modern societies, as noted by E. A. Ross, typically have a structure of opportunities for evil-doing different from those that prevailed in earlier societies.[53] Considering the extensive division of labor fostered by industrialism, Ross was apprehensive about the increasing opportunities for predatory and parasitic relations within our society, exclaiming, "Under our present manner of living, how many of my vital interests I must intrust to others!" Others who *may* betray that trust.

Because interdependence puts us "at one another's mercy," as Ross said, many new forms of wrong-doing become possible. Businessmen who profess social responsibility confront clienteles who are tempted to discount pious claims as mere propaganda ploys for maintaining public acquiescence in exploitative business practices. "Part of this skepticism," wrote the authors of a 1954 book on *Ethics In a Business Society*, "undoubtedly stems from the phoniness of the public relations conducted by some businesses," and from the fact that the advertising industry has become such a major component of the modern business system. "The citizen knows that a great part of the advertising campaign conducted for his purchasing dollar is based on an attempt to sell without very much regard for the truth."[54]

As skilled con-men presumably would realize, two characteristics facilitate successful exploitative efforts to misdefine the situation. If a hoax supports questionable but cherished beliefs and identities, it is more likely to be believed. If it supports local pride or patriotism, that also helps gain its acceptance.

But how could deception have become a normal aspect of human interaction? And how could *self*-deception arise?

Language and Lies in Hominid Evolution

Our ape ancestors involved in competition with other apes would have had reason sometimes to practice deception. One of the chimpanzees at the Gombe Field station provided a modern demonstration of this.[55] He had acquired an ability to open locked banana boxes. But he seemed to know it was unwise for him to do so in the presence of other more socially dominant apes who might attack him and take the bananas. To solve the problem this ape perfected the *acted* lie. By striding purposefully away from camp as if on his way to a good food source, he tricked other apes who would amble after him for a few hundred yards. By doubling back alone to the then deserted camp, he could open a banana box and peacefully enjoy its contents in the absence of the other chimps who, having seen there was no food in camp other than what was confined to boxes they could not open, did not return with him.

After the earliest protohumans became differentiated from apes, their hunting still depended on guile and deception. They used such ruses as the ambush and the covered pit-fall trap. Our existence indicates that our ancestors had successfully outcompeted neighboring groups of hominids, both ecologically and culturally, and left descendants.[56] Where feasible, the challenge was to dominate or exterminate their rivals. Deception would have continued to play a part as cultural advance by one group created selection pressure favoring comparable cultural advances among its hominid competitors.

Intra-species competition and aggression provided a context in hominid evolution within which there were real advantages to the development of language. As important in the evolutionary context as the need to convey clear and accurate information though, there was often a need to give false information, or confusing information, or actually to conceal information, and language could serve these methods for manipulating others.

To be effective a person's language must sufficiently overlap that of other members of a group to enable information exchange. The greater the social difference between speaker and hearer, the less meaningful the words. In the case of enemies, the less they know about one's relations, plans and activities, the less danger they usually constitute. Languages that diversify and constantly change afford an adaptive advantage by using new in-words to maintain the in-group's exclusivity. Language enables us to hide devious underlying thoughts. Appearing friendly can pave the way to eventual hostility. Promises not meant to be kept can be made. Respects not actually felt can be implied. Others' achievements can be exaggerated or belittled to raise false hopes or mollify suspicions.[57]

Written language was a logical development from speech and it apparently developed independently in several places on several occasions. It was far superior to word of mouth in enabling cultural information to be spread over a wide area and preserved through a long time span. Individual humans no longer derive cultural knowledge solely from their own experiences, or those of kin and close associates. Written language has made it possible to learn from many preceding generations—even from other cultures.[58] But writing, like speech, could also be used to create fiction and to deceive. And the cultural heritage from preceding generations may malfunction as changing conditions render traditional definitions of situations obsolete.

Deception as a Way of Life

Now, to escape the blinders of anthropocentrism, we must acknowledge that deceptive strategies that help prolong life, as well as straightforward honest strategies, have evolved among many species. Animals may be camouflaged by a coloration which matches their surroundings, or, they may mimic another type of animal by a combination of shape and color. "Deception as a way of life" is recognized by zoologists as something that "occurs throughout the animal kingdom."[59] Like any other adaptation, deceptive traits or actions result from—and are maintained by—natural selection.

Camouflage and mimicry function as both defensive and offensive strategies. Predators deceive to gain access to prey. But prey animals have evolved ways of deceiving their predators. Some plants, too, have evolved usefully deceptive odors or appearances. Deception is among the set of strategies by which an individual organism's chances of survival are enhanced, enabling its attributes to be transmitted to the next generation.[60]

Some birds can and do lie. Thrushes, for example, have learned to chase competitors into hiding by emitting their warning call signifying the presence of aerial predators. This gives them an opportunity to devour various tidbits undisturbed by rivals. Such avian behavior indicates lying is not something peculiar to humans.

It is because organisms coexist in a given environment and affect each other's life chances that deceptive appearances and actions are bound to evolve. Insofar as it is advantageous for an organism to avoid contact with, say, a noxious or harmful species, it may be useful to a harmless species to be mistaken for the one that is noxious or harmful. Mimicry has evolved because it induces in the signal-receiver a definition of the situation that is useful to the mimic. The fact that this is, objectively, a misdefinition of the situation (a lie) can sometimes put at risk its utility to the deceiver (the mimic), but only under certain conditions—such as too-frequent occurrence of too many mimics.

The Cuckoo Mimicry Dilemma

Several species of birds known as cuckoos are *brood parasites*, meaning they lay their eggs in the nests of non-cuckoos and thereby evade costs entailed by the task of feeding their own young. Selective pressure by the host birds (able to destroy recognizably foreign eggs)[61] strongly favors cuckoos that lay eggs exactly imitating eggs of the host species. A survey of known cases showed that only 8 per cent of 1642 cuckoo eggs laid in nests of the correct (egg matched) host were lost, while 24 per cent of 298 laid in wrong nests were lost. That is natural selection, differential reproductive success.

But success can become failure. This reproductive tactic of cuckoos can boomerang. Each female cuckoo needs several nests of the same host species for her eggs. Since the young cuckoo usually kills its non-cuckoo nest mates, a host bird fails to fledge progeny of its own when it successfully raises a cuckoo. While a poorly adapted cuckoo produces few offspring, a too well adapted cuckoo would inadvertently reduce the host bird population, to the detriment of its own future reproduction. By excessively harming their host species, parasitic species which are too successful undermine the deceptive practice upon which their support depends.[62] Continual and opposite oscillations of host and cuckoo population densities would presumably result from selection in both directions.[63]

These birds could have served as examples for Machiavelli (had he known about them). He held that political arrangements should be judged not by objective standards of right and wrong, but only by the law of success and failure. That view appears today almost unchallenged in business and politics.[64] In Chapter 18 of The Prince, on the basis of what is most helpful in political combat, Machiavelli discussed whether honesty and fidelity are good, and taught that a promise should not be kept when doing so would diminish one's own power (and when conditions are no longer those that occasioned the promise).

Machiavelli asserted the necessity of knowing how to be a "great simulator and dissimulator." Men were, he thought, so simple and so obedient to present necessities that a leader who would deceive "will always find someone who will let himself be deceived." Accordingly, a prince need not actually have all the virtuous qualities, but need only "appear to have them."[65]

Need to Deceive Obviated

The practice of sometimes intentionally misinforming others is deeply rooted in our culture—and may well be just as integral to *all* human cultures. In evolving as a language-using species our ancestors were acquiring enlarged abilities to behave deceitfully when the occasion seemed to call for it—and they also honed the ability to *be* deceived.

Is reduction of the need to deceive always good?

Not always. Consider the circumstances in 1942 shortly after U.S. declaration of war following Japan's aerial attack on Pearl Harbor. Some of those who pressed for having persons of Japanese descent (American-born and resident aliens alike) evacuated from the West Coast deemed it unnecessary to resort to euphemism. They explicitly advocated placing people who could be at least superficially identified with the enemy in "concentration camps." That ominous term was used even by the commanding general of the Western Defense Command, who finally made the fateful decision that sent thousands to what were subsequently called simply *relocation centers*.[66] Nor did all advocates of the relocation pretend national security motivation. The Grower-Shipper Vegetable Association, a California organization of white farmers, said about their reasons for favoring the 1942 relocation of persons of Japanese descent, "We might as well be honest . . . . It's a question of whether the white man lives on the Pacific Coast or the brown man. They came into this valley to work and they stayed to take over."[67]

What sort of conditions might diminish the need to deceive? When do ulterior motives or unjustified animosities *not* have to be disguised?

The persuasive potency of an idea may not be highly correlated with factuality. Study of the relocation episode suggests "emotionality" had a stronger influence than any facts. But what makes one idea more "emotional" than another? Morton Grodzins suggested a partial answer by pointing out that the war provided the unique situation whereby economic, racial, and political considerations could be confounded with patriotism (i.e., a desire to protect the American West Coast from the enemy). Perhaps, then, the general hypothesis could be set forth, stating that an idea will vary in potency in direct relation to the number of interests for which it serves as an abstraction. According to Grodzins, opponents of evacuation who rejected the racial basis for it "suffered not from a lack of facts but from the simple lack of a believing audience."[68]

One hopes no nation will ever again so unfairly mistreat, as a group, people of a particular ancestry as was done in 1942, but few are likely to deny that history *could* repeat itself in more abstract terms, and the possibility of any sort of repetition demands, because of its serious implications for higher human values, our careful consideration. Grodzins stressed that only one argument was repeatedly given in support of court enforcement of the evacuation policy: It was "recommended and considered essential by military authorities." In Congressional committee hearings, in committee reports, and in discussions before both houses of Congress, the key argument was that mass evacuation was "essential to the national defense." Not one congressman or senator investigated the basis for, or reasonableness of, that argument. And even those who expressed moral or political repugnance toward treating groups en masse qualified their objections by saying "unless required by the military"[69]

Conclusions

From the ubiquity of deceptive tactics in nature as well as in human society we can infer at least the following principles:

1) A given organism of a particular species has an interest in influencing other organisms' definitions of situations insofar as the given organism depends for survival or goal-attainment on the actions of other organisms (whether the other organisms are its conspecifics whose cooperation may be useful, or belong to a different species—e.g., predators to be avoided, or prey to be caught and consumed).

2) Deception consists of perpetrating a misleading definition of a situation. Signals that deceive may be elements of behavior (including human speech) but they may also be elements of an organism's "appearance" (in any relevant sense modality—sight, sound, odor) to another organism, the one that is deceived thereby.

3) Deceptive traits or practices result from evolution, or learning, in the on-going processes of interaction among interdependent organisms.[70]

4) Instead of being content merely to deplore deception because we moralistically suppose that honesty is always better than dishonesty, we should recognize that dishonesty (of which mimicry or camouflage are examples) *will tend to occur* when it appears to contribute more effectively to survival and propagation than would a non-deceptive alternative.

5) If modern division of labor enhances interdependence among a human society's members, it thereby increases both the opportunities for deception and temptations or pressures to deceive.

~

If we want to ensure honesty will be more prevalent than dishonesty, a taboo against deception, together with sanctions to "enforce" that taboo, are not likely to suffice. A society in which deception becomes rare must be a society in which circumstances make deception *less effective* than non-deceptive methods of goal-pursuit. In other words, when accurate definitions of situations are conducive to the well-being of all parties to the interaction, traits or practices conducive to misdefinition will be unlikely to occur and will not be reinforced or selected for. Just as natural mimicry loses efficacy when the ratio of mimics to models becomes excessive, lies of any kind (human or otherwise) become unavailing when they occur so frequently that they erode trust.[71]

Note that "*appears* to contribute more effectively" is the operative phrase in the fourth of the conclusions above. Appearances can mislead. Appearances easily convinced just about everyone until the time of Copernicus that the sun revolves around the earth, not vice versa. In thinking about the prevalence of deception we must never forget that there are occurrences of *self* deception, not always just due to plausible and non-deliberate misperception.

Familiar ways of living may *appear* sustainable even when they are imposing impacts that will bring fundamental change upon surroundings that *appear* permanent. Even "the everlasting hills" are not everlasting.

CHAPTER 4

How Specialization Reshaped Human Life

"the very different genius which appears to distinguish men of different professions . . . is not upon many occasions so much the cause as the effect of the division of labor."

—Adam Smith, *The Wealth of Nations*

"Give me your tired, your poor, Your huddled masses yearning to breathe free, The wretched refuse of your teeming shore, Send these, the homeless, tempest-tossed, to me: I lift my lamp beside the golden door."

—Emma Lazarus, "The New Colossus"
(Inscription for the Statue of Liberty, New York Harbor).

The use of different tools, adherence to different customs, and communication by means of different sets of symbols are some of the ways by which subgroups within a large human population can function *as if* they were members of different species. "Quasi-speciation" is a term I used in my book *Overshoot* to refer to this diversity of specialized subgroups.[72]

The culture-producing nature of *Homo sapiens* is what has made quasi-speciation possible. Among the members of a human population of any considerable size, in fact, our sociocultural nature has made differentiation *inevitable*. Without this aspect of humanness, we could never have formed complex societies, but one of the effects of modern division of labor, I believe, has been to alienate us more and more from one another.

If that is so, then what has made complex societies possible may have also made them self-destructive. Do we actually identify ourselves as all one species (which we still are, biologically), or do we behave as if we regard ourselves as so unlike other human beings that they are almost equivalent to being something other than human? Does modern life make it easier or harder to adhere to the religious precept of universal human brotherhood?

In the modern world, in addition to ethnic, racial, nationality, political, religious, social class and gender components of identity that distinguish us from one another, our individual concepts of "who am I?" depend to a remarkable extent upon our occupations. The different jobs of different people become a major aspect of their identities, of who (and what) they take themselves to be.

Most people do not understand either the obligations or the perquisites of most of those *other* occupations of the many people we encounter. Both the fact that each of us derives identity from our job and the fact that we have little understanding of most other jobs are characteristic of modern societies.

Specialization in extremis

Occupational specialization is a hallmark of today's societal life. Look at the last dozen or more pages in the back of a typical issue of the weekly magazine *Science* and you will see many columns of "Positions Open" ads placed by colleges, universities, government agencies, industrial corporations. Most begin by saying something like "The Department of ____ at ____ invites applications for appointment to the position of ____"—followed by some indication of the educational background required, the general duties, the salary, etc. No matter how highly qualified you may be for *some* occupation, you will surely find yourself *unqualified* for the vast majority of the advertised positions, each of which you will recognize as being suitable only for someone else. And you will usually have at best only a vague idea of the actual tasks a particular ad implies and the particular skills required in any of the positions alien to your own qualifications.[73] Yet you and the successful applicants for each of those jobs are all members of one species—*Homo sapiens*.

I was reminded of how intricate the division of labor can become in modern societies on a Sunday morning shortly before the breakup of the Soviet Union as I watched the ABC-TV program "This Week with David Brinkley." That day it included interviews with William Hyland, Editor of *Foreign Affairs*, Georgi Arbatov, member of the USSR Congress of People's Deputies, and George Schultz, former US Secretary of State, followed as usual by discussion among the program's four regulars: veteran journalists David Brinkley, Sam Donaldson,

Cokie Roberts, and George Will. But that particular Sunday's program ended in an extraordinary fashion. Instead of the usual rather cynical homily about political foibles with which Brinkley normally closed out the hour, he said this [my italics]:

> Finally, for George, Sam, Cokie, and all of us, I want to wish you the very best for Christmas, the holidays, and the new year. All of those who are with us every week, and all four of us offer our thanks to *a long list of people whom you at home do not see or hear but without whom you would not see or hear any of us*—all of those ABCers behind the scenes. And here they are; now pay attention. Stay there and pay attention.

Then, on the screen two sponsoring corporate names appeared (Merrill Lynch and General Electric) followed by the rolling of a surprisingly long list of position titles—47 of them—with one or more names of persons beneath each title.

A total of 156 unseen persons were revealed as being at least somewhat involved in a TV program which, to the viewing public, had the appearance each week of involving just the four on-screen "regulars" plus two or three interviewees, plus one or two reporters in whatever "stand-ups" and "backgrounders" were shown in the program's early moments. More importantly, these unseen persons performed *47 variously specialized roles*, all essential in one way or another to providing viewers access to the familiar faces on screen and to thoughts from the brains behind those faces.

The same kind of indication of behind-the-scene involvement of myriad specialist occupations can be obtained at a movie theater by simply sitting through the rolling of a film's "credits." An impressive number of people exercising an impressive number of differentiated skills will be seen to have made possible the images of actors on the screen.

Because occupations in modern societies are both specialized and numerous they are one of the principal ways in which we, the mass of people, are differentiated from each other. Specialties do confer identity. We all sense this, so much so that it is even reflected in idle conversation. Any two unacquainted individuals who happen to encounter each other in a queue—for instance at the passenger check-in counter of a major airline, are unlikely to be employed in identical jobs. Should they find themselves in adjacent seats aboard their flight and begin to get somewhat acquainted through conversation before or after takeoff, some approximation of "What do you do for a living?" will be one of the early questions—probably subsequent by only a moment or two to the even more stereotypical "Where are you from?"

The enormously ramified division of labor in modern societies is one of the factors responsible for the fact that weather has become one of the few topics each of two strangers randomly confronting one another can be sure is a mutual interest, about which both are likely to have opinions. Unless one of them just happens to be a meteorologist or a climatologist, however, their respective (non-expert) opinions about the subject are unlikely to have any significance beyond enabling a friendly exchange of remarks to occur that may be socially preferable to the awkwardness of prolonged silence.

So our occupational differences set us apart from each other. The fact that different vocabularies go with different skills and create communication barriers is just one of the ways in which human beings are divided from one another—in addition to whatever ethnic, educational, marital, political, racial, religious, or recreational differences may separate us.

The Significance of Specialization

How did specialization *arise* in human societies? The English social philosopher Herbert Spencer regarded population growth and the resulting pressure upon scarce resources as causal for growing specialization, but Emile Durkheim, in a book that was to become a sociological classic, elaborated upon the idea to suggest the pressure was exerted *in two phases*. The first phase he called increased "material density," a rise in the sheer number of people struggling for livelihood in a given place. The second phase, upon which Durkheim's attention was focused, was an increase in "moral density," the multiplying *numbers of encounters* between individuals as their material density grew. Life would become more complicated, perhaps insufferably more so, as numbers grew. Specialization, he said, was a means of limiting the number of encounters that for any particular individual had to be salient.

An additional important effect of specialization is what a modern ecologist would call "resource partitioning."[74] Each worker, according to his or her occupational category, earns a particular claim (of a market-determined magnitude) on the array of products that results from the several lines of work by others. Dividing labor along specialist lines allocates entitlements of labor force members to the environment's resources and the products of labor. These quantitative entitlements-to-purchase are expressed in money, of course, and in a later chapter we will consider some of the ways in which monetary aspects of life affect human identities.

Spencer had invoked biological division of function as a basis for understanding human division of labor. Across the Channel, in France, Durkheim likewise supported his own argument by reference to parallels in biology. In particular, he cited and quoted Darwin, for he was writing when the whole learned world had had several decades exposure to Darwin's emphasis on biological diversity and the effects of natural selection.[75]

So we see how full understanding and appraisal of human division of labor requires some familiarity with Darwinian ideas and an awareness of their subsequent development.

Relevance of Evolution and Ecology

One effect of publication of Darwin's *On the Origin of Species* was to enable students of evolution to get beyond the question "Does evolution occur?" and focus instead upon perfecting knowledge of the processes by which it occurs. By the Darwin centenary in 1959 there had accumulated, of course, many relevant scientific achievements not yet available when Darwin worked out his theory, and it was becoming clear that advancement of evolutionary knowledge "rests largely with those who combine the techniques and viewpoints of genetics and ecology."[76]

Here, then, is a thumbnail summary of Darwinian evolutionary process as understood a hundred years after Darwin presented his theory of natural selection to the world[77]:

> Today we know that any widely distributed species population is *broken up into many subpopulations,* each with its own pool of adaptive hereditary characters. Each pool is the result of *selection by the local environment for characters that are adaptive to that environment.* Variability is provided by gene mutations and chromosomal changes and by the tendency of individuals to disperse through the area occupied by the species. The tendency to vary is countered by the selective elimination of the least fitted in each of the many local environments in which the species lives. The least adapted leave fewer offspring and the best adapted leave more offspring and so predominate in succeeding generations.

For comparison, here is a paragraph from Spencer's *First Principles*[78] as quoted by Durkheim[79]—who intended to supersede the theory expressed by these words of Spencer:

> A community which, growing populous, has overspread large tract, and has become so far settled that its members live and die in their respective districts, keeps its several sections in different physical circumstances; and then they no longer remain alike in their occupations. Those who live dispersed continue to hunt or

cultivate the earth; those who spread to the sea-shore fall into maritime occupations; while the inhabitants of some spot chosen for its centrality, as one of periodic assemblage, become traders, and a town springs up . . . . A result of differences in soil and climate, is that the rural inhabitants in different parts of the kingdom have their occupations partially specialized; and become respectively distinguished as chiefly producing cattle, or sheep, or wheat.

Clearly, Spencer's explanation of specialization among humans closely paralleled the Darwinian explanation of *speciation* (the separation of a population of organisms into differing descendant populations).

Durkheim, though, was unwilling to rest with Spencer's view of the matter. "Even when we compare, not functions very remote from one another, but only various branches of the same function," wrote Durkheim, it is often completely impossible to perceive to what external dissimilarities their separation from one another can be ascribed." To nail down his point, he became more specific[80]:

Scientific work is continually being more divided up. What are the climatic, geological or even social conditions that can have given rise to those very different talents possessed by the mathematician, the chemist, the naturalist, the psychologist, etc.?

Certainly his point seems well taken; specialization in science, if it is to be accounted for by environmental influences, must take into account something other than diverse *biophysical* environments.[81]

In further development of the theory of Darwinian evolution since Darwin's time the emphasis has been on accounting for speciation by noting how variations among different local environments cause diversity in the selection pressures they exert. Speciation results when the contrast between segments of a diversified population is sufficiently increased by their exposure to different selection pressures that they become no longer able to interbreed. This process turns diverse subpopulations' descendants into separate species populations. This is more than just a matter of *resource partitioning*. Resource partitioning refers to the way in which the various species in association within a local ecosystem are enabled to co-exist by being sufficiently differentiated from each other in their respective niche requirements.[82] Speciation achieves such differentiation of niche requirements—and goes beyond it as the differences become barriers to interbreeding.

But it was that idea, *co-existence by niche diversification*, that had caught Emile Durkheim's attention and which he used to develop his insights into human division of labor. He inferred quite plausibly that occupational specialization limits competition. People compete with others in the same occupation, but much less so (if at all) with those practicing different occupations. To support this view, Durkheim leaned heavily on Darwinian observations about coexistence among diverse species. If different species make different demands on their shared environment and thus do not directly compete with each other, wouldn't this likewise hold for different occupations?

But he was reaching too far in supposing co-existence *caused* niche diversification. What does actually *cause* species diversity? Does diversity result, as Durkheim assumed when he sought to draw a human parallel, from a need to minimize competitive pressure?

In the tropics, according to a number of sources cited by Peter Price,[83] there are many species packed together within small areas. Each species exploits a different niche—*as if* interspecific competition had forced them all to specialize. By contrast, in regions farther from the equator, ecological communities tend to be more loosely packed. Competition between the species populations within them seems less severe.[84] This *seems* like a pattern Durkheim would have expected.

However, it is important to consider not just the fact that species packing varies among different places, but the fact also that in this case the places differ in other respects. *Why* are there more species in tropical ecosystems and fewer species in temperate ecosystems? What is it about the tropical vs. temperate contrast that makes a difference?

It turns out that the coexistence of greater numbers of species packed into tropical ecosystems is not so much due to any greater specialization among them but can be attributed to the presence in tropical locales of a greater diversity of resources. Instead of specialization apparently resulting from intensified competition for scarce resources, it seems to occur where there is a greater *variety of resources* to be exploited.[85] So it could almost have been Durkheim's sociological theory from which non-sociologist Price was dissenting when he concluded on this basis that "the real world differs substantially from theory on community organization."

From Competition to Mutualism (Organic Solidarity)

Durkheim's next theorizing step was to argue that once the competitive aspect of the struggle for a living is minimized (supposedly by diversification), the *interdependence* that must result from specialization will become a new form of social solidarity. What he was suggesting can be most clearly expressed by further embracing a vocabulary not yet available in his time—the vocabulary of modern ecology.

Competition for scarce resources by increasingly numerous competitors was assumed by both Spencer and Durkheim to be the force behind the division of labor. But competition is by no means the only relationship occurring between living organisms sharing a common locale.[86] If there has been overemphasis on selected antagonistic relationships—predation and competition—it may represent misunderstanding of Darwin's influence upon subsequent ecological thought.[87]

Division of labor in human society, Durkheim believed, is a special human instance of a pattern one finds everywhere throughout the living world. The pattern he had in mind was what ecologists now call symbiosis, the living together of unlike organisms. When their differences serve each other's needs and make them interdependent, mutualism is the ecologists' term for their relation. For ecologists, the word mutualism means the mutual dependence of populations of differentiated organisms.[88] (In the vocabulary of ecologists, symbiosis just means "living together," and can refer to *any* of the relations occurring among organisms that co-exist within a given environment, including competition, predation and parasitism. But the word symbiosis is sometimes used as if it were a synonym for mutualism.) Mutualism is the *particular kind* of symbiosis that appealed to Durkheim and he chose to call it "organic solidarity." His term is still used by sociologists.

In the human instance, division of labor would limit the range of competition, result in interdependence among different specialists, and produce a mutualistic relation, Durkheim's "organic solidarity." For Durkheim, it seemed as if specialization overcame the competitive relation and *enabled* mutualism to develop. But if his analysis were fully valid, today's most modern societies should have very strong organic solidarity.[89]

Differentiation Aggravated

What prompted my writing this book is my conviction that events of our time severely challenge the view that enhanced social solidarity has been the actual result from the occupational specialization wrought by population pressure and technological advances, especially all the amazing advances since Durkheim wrote. As I follow the news from day to day and week to week, I see many indications that diversity is associated with the very *opposite* of solidarity.

Adherents even of religions that extol universal human kinship continue finding it easier to regard people who belong to other faiths as enemies. People who embrace alien belief systems tend not to be respected as pursuers of alternative human aspiration but are condemned as infidels.

Foreigners are too often deemed less worthy than fellow-citizens of being treated according to our own nation's conception of civil rights. It is tempting to label anyone critical of our values and practices, especially if suspected of thinking about acting violently in opposition to what are thought to be our own country's interests, as a "terrorist"—that word having come to connote someone not quite human.

Racism has not yet become extinct, though some of its past extremes have become less common, at least in some places. Racial discrimination in the job market has not yet been eradicated.

Antagonism by in-group members toward out-groups thus has to be recognized as an old story, but one thing that makes it relevant to the issue of ultimate societal effects of modern division of labor is the way inhabitants of lands where there happen to be abundant quantities of natural resources required to support "our way of life" tend to be seen not as the rightful beneficiaries of that wealth, but either as troublesome impediments to *our* access to it, or, at best, as a potential labor force to be employed in the process of providing to us "the oil we need" etc.

And a glaring ambivalence became apparent as politicians debated proposed legislation to "reform" American immigration policy. Most non-Indian Americans are descended from people who came to this continent when it was less crowded and when "immigration" was restricted only by nature's impediments to travel and not yet by xenophobic laws enacted by human governments. Even among people with several generations of American-born ancestors, their first arriving forebears came to the North American continent from elsewhere *without a visa*. Before there were immigration laws there were no visas and no *illegal* immigrants.

The phrase "undocumented immigrants" is today a dehumanizing label too glibly applied to an estimated twelve million foreign-born persons residing in the U.S. To be sure, no one seems yet to be proposing to chisel away the famously hospitable words of Emma Lazarus on the plaque at the base of the Statue of Liberty, and people everywhere else are still assumed to have a universal urge to immigrate to "the land of opportunity." Nevertheless, as a proposed solution to "illegal immigration" establishment of a "guest worker" program has been suggested. This would "welcome" entrants from other lands (but only temporarily—not as people but as conveniently impermanent components of a labor force). Much political discourse depicts certain human beings as mere tools, temporarily admissible only insofar as they would retain their utterly instrumental status.

One ought to wonder how these attitudes about other human beings must appear to those others. People ought to consider how that appearance of regarding others as mere tools might influence the attitudes of those others toward them. More generally, when the world's population of *Homo sapiens* is defined not just as "people" but as "peoples" the thought ways this implies hardly accord with any optimistic notion of "organic solidarity." Any in-group is to its out-groups an out-group. Hostile (or merely wary) attitudes of in-group members toward out-groups should be expected to influence out-group members' attitudes toward that in-group. People can be expected to reciprocate the apparent hostility or wariness of others.

CHAPTER 5

How We Misconstrued Human Division of Labor

"How can the poor and disenfranchised have their voices heard when they express outrage at the unequal share of the burden of industrial pollution their countries and communities have had to bear?"

—Gerald Markowitz and David Rosner,
Deceit and Denial, p. 2.

"Which particular tasks and occupations are defined as 'men's' and which as 'women's' will vary according to time and place. . . . In all cases for which evidence exists it points to a sexual division of labour in pre-industrial societies."

—Harriet Bradley, *Men's Work, Women's Work*, pp. 2, 223.

What is it that really troubles the world today?

As the speaking species we have elaborated fantastic technologies and potent forms of social organization. As I have argued elsewhere[90] the most technologically advanced populations of human beings are no longer simply *Homo sapiens*. The technological apparatus at our disposal enables us to behave as giants. In view of our now prodigal resource appetites and tremendous environmental impacts we have become *Homo colossus*. Without language (and other symbol-wielding abilities) we could not have thus transformed ourselves and would not have grown so numerous nor have so drastically altered the role of our species in local ecosystems or in the biosphere considered globally. Are we hell-bent on a course of *cultural* evolution and demographic expansion that will so modify this planet that we must find we have deprived *our own species* of suitable habitat?

Having evolved as the species with language we can excel all other species in formulating definitions of the situations to which we respond. But with language we can also *mis*define. Our language ability includes no automatic guarantee that we will speak wisely, no automatic guarantee that our word-maps will always correspond to external reality.

We live in a time of pervasive foreboding. But a valid, conclusive diagnosis of what ails us has so far eluded most leaders in government and business, most members of the general public, and most sociologists. Of course there is growing awareness that our future is imperiled by such things as climate change, depletion of natural resources, environmental degradation, pandemic disease, etc., and there is genuine apprehension that terrorism will be with us far into the future. There is dread that terrorists or "rogue nations" might acquire devastating weapons of mass destruction.

But why did grave concerns like these come to dominate the thoughts of humanity when not long ago we believed in limitless possibilities? Such hubris was symbolized and reinforced by emphatic defeat of totalitarian aggressors in the Second World War, impressive postwar recovery, and then the marvels of dawning space age.

Competition, Specialization, Solidarity

The basis for today's anxieties has been brewing for a long time. Language and culture enable human beings to become differentiated from each other in many ways—by nationality, ethnic identity, religious persuasion, and especially nowadays by occupational specialty. But differentiation among living beings long preceded human language and its many influences. In the biosphere and in localized ecosystems there is division of function between various associated species of organisms.

That fact caught the attention of Durkheim. According to the theory he expounded in his influential 1893 book, written just over one century into the industrial era—the *abatement of competition* and the rise of *organic solidarity* were two major effects of the European level of division of labor in that first century of industrialism.

Human occupational differentiation seemed to Durkheim a special case of a general biological process of speciation—a kind of quasi-speciation (though he did not have that term to apply to it). But he did think actual speciation had

49

for other members of the animal kingdom the effects he supposed occupational specialization was having within European societies.[91]

His view on this was based on ideas he thought he had drawn from Darwin. If he had somewhat misread Darwin, as had many of his contemporaries, *his* misreading was different. He never advocated so-called "social Darwinist" ideas—notions that distorted Darwin's concept of natural selection to glorify cutthroat competition in the business world.

Durkheim seemed to suppose speciation occurs as a means of enabling different species to avoid or minimize competing with one another. However, writing when he did, even this brilliant and hard-working pioneer sociologist could not yet know enough about biological evolution to understand that (in modern evolutionary terminology) speciation was not generally *sympatric*; it was usually *allopatric*. (See glossary.) In plain English—among species which are separated geographically from each other (i.e. allopatric species) it is their difference in location, not their differences in form and function, that precludes competition between them. They aren't in competition with each other simply because they are not in the same place, and thus not having to use the same environment, not confronting the same resource limits, etc. Each species has adapted to using resources that were available in its local environment, regardless whether the array of resources available to its sibling species elsewhere in other environments was the same or different. For abatement of competition to be "the function" of speciation, speciation would have to occur *sympatrically* (i.e., without geographic separation), but that is apparently not the usual way it happens, a fact unknown to Durkheim.

In sum, unless the several different daughter species arose typically within the same local environment inhabited by an ancestral species from which they all descended, there would not have been any immediate opportunity for them to compete with each other. Evolving in different places, without any need for competition between them to be softened, competition-diminishing differentiation could not be the particular survival-enhancing benefit implicit in the selection pressures to which the daughter species circumstances subjected them.

In addition, biologists' studies since Durkheim's time have shown that what ecologists call *mutualism* (the interdependence of differently specialized organisms that Durkheim called "organic solidarity") doesn't typically arise in the way Durkheim supposed. Mutualism is not generally a relationship between former competitors who have specialized so as to reduce competition. Mutualistic relations may instead arise between members of a predator-prey pair when the prey species evolves in some way that changes the outcome of the pair's interaction.[92]

Mindful of what we learned in Chapter 2 from Carl Becker's study of idea-development in the writing of the Declaration of Independence, to assess where Durkheim's ideas about division of labor, written more than a century ago, would lead us today, let us consider why he wrote what he did. What he wrote was shaped partly as a rejoinder to earlier notions about the subject.

Forerunners of Durkheim

Division of labor was not a new concept in Durkheim's time. Nor was it simply a new characteristic of society born of the Industrial Revolution. Although Durkheim put a spotlight on the *greater degree* of occupational specialization, differentiation, and coordination that arose as societies became industrial, the concept of division of labor among human members of a society had been undergoing gestation through many centuries.

For Plato, division of labor meant a three-part social structure. There was: (1) rule by the wisest, a hereditary king, with help from (2) society's most courageous members in maintaining order and defending the state, while (3) others performed a variety of mundane tasks.[93]

After the concept's prolonged sojourn in philosophy's womb,[94] human division of labor emerged to be seen by Adam Smith[95] as a means of economic progress—an arrangement enabling a work force to produce more goods more efficiently than would be produced in "that rude state of nature" where every individual "endeavors to supply by his own industry his occasional wants as they occur."

Smith had illustrated specialization's advantages by taking several pages to describe the manufacture of pins by a work force of specialists, each performing a single type of operation in the production process, thereby collectively turning out far more pins per unit of effort. Of course, a group of pin-makers would thus turn out far more pins than *they* would ever need for themselves. They would do so only because, through sale in the market, pins could be the means of acquiring many other wanted things—produced by other, differently specialized, workers.

This quantitative increase in output from a given number of workers Smith attributed to three different circumstances[96]:

> first, to the increase of dexterity in every particular workman; secondly, to the saving of the time which is commonly lost in passing from one species of work to another; and lastly, to the invention of a great number of machines which facilitate and abridge labour, and enable one man to do the work of many.

From Smith's time onward, such advantages from division of labor have been axiomatic in economic thinking.[97] Durkheim, whose emerging academic identity as a sociologist separated him from "the economists" (a phrase he used somewhat pejoratively), looked at the matter from a different angle and saw, among division of labor's consequences, *interdependence*—as a more significant effect, he argued, than increased productivity.

Our angle of vision today differs from both Smith's and Durkheim's, and calls into question Durkheim's evaluation of that interdependence effect. We can also see the matter differently from Durkheim's English contemporary, Herbert Spencer, a writer with a broad range of interests, an ardent proponent of a generalized evolutionary view of change in all realms from cosmology to sociology. Spencer believed progress essentially consists of the transformation of the homogeneous into the heterogeneous.[98] He correctly considered division of labor not only as an economic but also as a biological phenomenon, and suggested its relevance for the wider study of social evolution.

There had been division of "labor" in the biotic world long before there were humans. Already in the plant kingdom there is sex—which means a division of the tasks that achieve reproduction and perpetuate a species. In the organic world at large, there are *autotrophs* and *heterotrophs* and *decomposers*. The autotrophs can produce their own "food" by synthesizing organic molecules from inorganic materials using energy from sunlight; whereas heterotrophs cannot do this, but subsist by consuming autotroph substance as food. The decomposers start nature's *recycling* of the substance of dead organisms (be they autotrophic or heterotrophic).

Even within single-celled organisms (as we know now) there are differentiated parts, and the unlike parts perform different tasks in the life processes of these microscopic creatures.

What Spencer had to say about the social significance of division of labor,[99] among many other matters, exerted a lasting influence upon what was then the very new field of study called sociology. Arguing there was a need for sociologists to have some grounding in biological science (a precept later unwisely neglected by several generations of sociologists), Spencer said biology had borrowed the division of labor concept from social thought and was giving it back to sociology,[100] "enriched by countless illustrations, and fit for extension in new directions."

Conception of "the physiological division of labour" was credited by Spencer to a Belgian-born Englishman, Milne-Edwards,[101] but Spencer said the idea "obviously originates from the generalization previously reached in Political Economy," which he stated as follows[102]:

Recognition of the advantages gained by a society when different groups of its members devote themselves to different industries, for which they acquire special aptitudes and surround themselves with special facilities, led to recognition of the advantages which an individual organism gains when parts of it, originally alike and having like activities, divide these activities among them; so that each taking a special kind of activity acquires a special fitness for it. But [in] Biology, this conception was forthwith greatly expanded . . . [and] found applicable to all functions whatever. It turned out that the arrangements of the entire organism, and not of the viscera alone, conform to this fundamental principle . . .

The idea, "thus developed into an all-embracing truth in Biology," said Spencer, "returns to Sociology ready to be for it, too, an all-embracing truth."

Emphasis on Interdependence

Durkheim was the kind of scholar inclined to be skeptical of any allegedly "all-embracing" truth, but the social significance of division of labor did indeed seem to him more profound when it was not construed merely as a means of increasing production and attaining efficiency. Nor did he regard it as caused just by the desire for greater material abundance. He focused upon quite a different aspect of Smith's analysis, the exchange of diverse goods and services among the various specialists. For Durkheim, as we saw in Chapter 4, it was the fact that specialists necessarily became inter-dependent by virtue of their diversity that seemed of paramount importance.[103]

I will argue in later chapters that because of further elaboration of division of labor in the 20th century—since Durkheim's classic analysis in the 1890s—the dysfunctional effects have come to outweigh what he claimed to have shown was its function, the fostering of social solidarity based on mutual interdependence among diverse specialists. By and large, sociologists have failed to see or understand this reversal, and continue to adhere to his benign expectation, as if there were no question that to whatever extent modern societies are cohesive this is substantially due to division of labor. They continue to regard his study of the division of labor in 19th century Europe as a classic, and believe it provided a seminal answer to the question What is the function of division of labor in society? So his concept of "organic solidarity" persists as the expected and ostensibly enduring product of interdependence among people in a system with proliferating diversity.

According to the perspective inherited from Durkheim, not unless what he called the "abnormal forms" of division of labor become excessive should societies begin coming to grief and falling apart. Compared to the rest of his argument, which was basically optimistic in what it presumed division of labor would do for humanity, too little space was allocated in his 1893 book to discussing the "abnormal forms"—hostility between labor and capital, involuntary servitude, idleness and wasteful work. But even normal interdependence induced by division of labor turns out to have seriously insidious consequences, as we will see in later chapters.

Signs of incipient disintegration in societies today should cause us to wonder why Durkheim's optimistic expectation has lost validity. Or was it valid in the first place? Was the theory about solidarity-by-interdependence actually ever empirically confirmed, or in any way explicitly tested? Should we really *expect* the degree of interdependence to which we are subjected by modern division of labor to "solidify" a modern society? If not, why not?

Was "organic solidarity" a mere fantasy?

A major distraction has effectively obscured the issue. Recently prevalent misuse of the very concept "division" of labor did this. Although Adam Smith and Emile Durkheim differed as to what advantage they took to be most significant, when they wrote about division of labor both meant *differentiation and specialization*. In recent decades, though, that Smith/Durkheim "debate"—over which advantage to emphasize—has somewhat faded out of public (and scholarly) awareness, derailed as writers slid into usage of the phrase "division of labor" in a fundamentally different way, having little to do with specialization and more to do with equitable or inequitable *sharing* of some particular task. Like "division of the spoils," which has to do with which thief gets how much of a common pile of loot, for certain interest groups division of labor meant who does *how much* of the work.[104] When we think of "division" as mere *sharing* of *one* task—a task that seems to be more or less a single kind of activity, then the issue we may trouble ourselves about is the question: "Is each individual doing his or her fair share of the work?"—not "Is each doing what he or she does best?"

We lose sight of aptitude diversity, or trained differentiation of skills or preferences. When we view division of labor as mere sharing of tedium, or of onerous effort, we are more inclined to see issues of equity and not issues of interdependence. So we fail to focus on the consequent vulnerability and exploitability that arise from specialization-induced interdependence. Equity (or fairness) can be a legitimate concern, but it is not the same issue as the allocation of different kinds of tasks among persons with different skills and interests, different training.[105]

Tom Sawyer's Version

Understandably, we have tended to look at life as did young Tom Sawyer in the story by Mark Twain about whitewashing a fence.[106] Tom hated to have to give up leisure time to a dull, tedious task his Aunt Polly had assigned him. However, when other boys came along and started to tease him for having to work while they were free to play, he was inspired to pretend the mundane work involved a skill he was proud to possess. By insisting his Aunt Polly was very particular that the job be done meticulously, he aroused envy in his teasers, and thus he teased them into paying him (with whatever items of boyhood value they had in their pockets) for the "privilege" of painting a portion of the fence—under his watchful supervision. So he divided the labor among several substitute (and subordinate) workers, eventually astonishing his Aunt Polly when he cheerfully announced to her the job's timely completion.

Of course, Tom Sawyer was a character in fiction. But real people can resonate with the story because we are all inclined sometimes to avoid drudgery and leave it to others if we can do so. Are we in fact likely to use the words "division of labor" to mean that sort of thing instead of what Smith and Durkheim meant? Sitting in the waiting room at an automobile dealership where I was having my car serviced, I actually happened to be passing the time by studying a then-new translation of Durkheim's book. When the manager of the business came by and made friendly inquiry, "What are you reading?" I replied in a very matter-of-fact tone of voice, hoping to avoid extended conversation so I could get on with my own thoughts: "It's a sociological classic entitled 'The Division of Labor in Society.'" Before strolling onward and leaving me to my own pursuits, this head of a local business responded in a way that indicated he had no idea what either Durkheim or I had in mind. "I wish I could divide *my* labor," he said.

At the time I construed his remark as indicating a wish for something akin to Tom Sawyer's re-allocation of fractions of his tedious task to others. Subsequently, though, I wondered if by "my labor" he had meant not his work but his workers. Was he implying something other than the wish for a reduced personal workload? Was he perhaps thinking in "divide and conquer" terms, wishing he could divide his *labor force* to weaken their collective power to resist or deviate from his commands?

But there are other instances in which people neglect the difference between subdivision of a task into uniform pieces and subdivision of a complex process into specialties.

Non-fiction Examples

To see this shift of meaning, consider baseball. In baseball there is a division of labor among the nine members playing for a baseball team. They have nine *different roles* on the field, and their coordinated actions are all essential to playing the game.

One of the rare occurrences in baseball is a "perfect game." The honor for such a performance is credited to a pitcher who works nine complete innings and faces only 27 batters—with no member of the opposing team reaching first base.[107] Achieving a perfect game does not mean that the pitcher strikes out 27 consecutive batters, much less that he throws exactly 27 X 3 = 81 pitches, all strikes.[108] He may throw fewer than 81, or possibly more than 81, some pitches being outside the strike zone, and perhaps some pitches getting "fouled off" after a batter's second strike. What matters is that no opposing batter reaches first base safely—either by a hit or a base-on-balls (or being hit by a pitch).

The perfection in such a game is of course achieved not just by the pitcher; it depends on essential assistance by the other eight players on the field who catch fly balls for outs, field ground balls successfully and throw quickly to first base for outs. Some (or even all) of the opposing hitters may make contact with the ball and "put it in play." If they pop up, and the ball is caught by any of the pitcher's eight teammates, so that the batter is out, the teammate fulfilling his particular role by making that catch has contributed to "the pitcher's" perfect game. Or, when a ground ball is hit to, say, the third baseman or the shortstop, who fields it and throws across the diamond to the first baseman in time to put the batter out, both the infielder scooping up the ball, and the first baseman catching his throw in time for the put-out have contributed to "the pitcher's" perfect game. All part of the division of labor, *in the original sense of the term.*

But there are anomalies in baseball and they can be illuminating. In 1917, when Babe Ruth was still a pitcher for the Boston Red Sox (before he achieved his principal fame as a home-run hitting outfielder for the New York Yankees), he was ejected from a game he started against the Washington Senators. He gave up a base-on-balls to the first batter but then complained too vociferously about the umpire's call on ball four. Umpires throw players out of the game for that, and Ruth was ordered off the field. Ernie Shore came on to replace him, pitched a full nine innings, and when no subsequent Washington player reached base, got credit for a perfect game—*with an asterisk in the record book.* It was only an *almost* perfect game. That lead-off Washington batter, after being walked by starting pitcher Babe Ruth, was thrown out trying to steal second. Shore faced not 27 batters but only 26, none of whom reached first base.[109]

The next perfect game in Red Sox history was also anomalous, in another way. It came nearly 83 years later during spring training, in a game against the Toronto Blue Jays. Because it was spring training, rather than a regular season game, six Boston pitchers saw action (and shared the perfect game credit). That is the anomaly, and the newspaper coverage of the game became relevant to our consideration of slippage on the division of labor concept. In the Sports Section story by Michael Madden for the *Boston Globe* newspaper, the table of six "pitching lines," showed 3 innings pitched by one man, 2 by another, and 1 each by four others. There were zeroes all the way down the columns for hits, runs, errors, and bases on balls, and a total of 11 strike-outs. The heading above this table of zeroes called it: "Perfect division of labor." And that is what makes it of interest here.[110]

That six-way allocation of the *single task*[111] of pitching is not the sort of divided-into-specialties effort that Durkheim meant by division of labor. That sharing of the essentially undifferentiated pitching job (between those six men, or 83 years earlier between Ruth and Shore) is not what the phrase meant to Adam Smith, or Herbert Spencer, or Karl Marx—nor to most sociologists. Until very recently. Then a new sociological interest in gender roles developed with the rise of modern feminism. A body of sociological literature on "household division of labor," so-called, emerged in which the issue was equitable versus inequitable *sharing* of essentially *identical* tasks by husband versus wife, rather than, a la Durkheim et al., the coordinated *differentiation* of task roles between diverse specialists.

The issue was commonly seen as "Does he do his fair share?" rather than "Does each perform a contributing specialty?" Fairness is, of course, a perfectly legitimate issue in various human relations, in families as in a sport like baseball, or in industry, commerce, or combat.[112] This focus on fairness questions became, however, a distraction from previous issues involved in the study of division of labor (in the diversity sense of the phrase).

Changed Circumstances and Changed Fairness Standards

Changing circumstances, though, change the manner of conceiving what is fair and proper, and what is to be deplored as unfair. Questions of fairness in the relation of gender roles acquired increasing salience as historical events caused societies to change, such that new societal conditions undermined old assumptions about relations between males and females.

Already in the nineteenth century, technological changes were bringing about employment opportunities for women as alternatives to such a housewife role as had been common among Europeans, and among Americans in colonial times. To some extent that influence was accelerated in the early twentieth century by the first World War. Wars have side-effects, and one side-effect of economic adjustments in that war was increasing employment opportunities for women. But afterward there was retrenchment during the Great Depression.[113]

Then a renewed and greater acceleration resulted as mobilization of the young adult male population for military service became extraordinarily extensive in World War II. Many civilian jobs formerly performed almost exclusively by men were opened to women, who were actively recruited to fill them. The de facto redefinition of "women's work" made earlier distinctions between "women's work" and "men's work" anachronistic.[114]

After the Second World War ended in 1945, demobilization led to the forming of many new families. Under new circumstances, and in the aftermath of the wartime experience, old assumptions about male-female relations within even the family context were increasingly disregarded or deliberately contested.

Toward Androgyny

In all human societies, people learn to play prescribed roles. The role expectations vary from one society to another, but in any society they change as we grow up, and in many societies, each individual may have numerous roles—relating differently to different others, in different contexts. Whatever else humans have become, however, we are still biological entities in part, and our biological characteristics remain among the array of factors in the assignment of roles—even though the evolution of human cultures has brought many other considerations into the shaping of role expectations. Much of the social and cultural turmoil of recent decades can be understood as a somewhat fumbling dissociation of non-reproductive role expectations from the reproductive roles of the two sexes.

One domain in which there has been remarkable gender dedifferentiation in recent decades is the military establishment.[115] This dedifferentiation is less than total, but its extent has been considerable, and it may afford clues as to the reason for an unDurkheimian (or Tom Sawyer-like) usage of the division of labor phrase in the studies of "gender equity." When we spotlight the occurrence of *de*differentiation, we may become distracted from considering some of the causes and consequences of differentiation.

Most societies for most of human history excluded women from military service. Women tended to be taken into the armed forces only when men were in short supply. In the Soviet Union during World War II, the situation was quite desperate, and women fought heroically. Under "normal" circumstances, however, women's limited use by military establishments reflected their restricted admission to other male-dominated occupations.

Each human society has had its characteristic perceptions of appropriate male and female roles. In the Jewish and Christian faiths, as well as in ancient Greece and Rome, men were taken to be protectors of the home. Women had been seen as nurturers and care givers within each home.[116] But with the onset of World War II there were abrupt changes in the roles taken on by women, both at home and in the military. American involvement in the first World War had been too brief for any such extensive change, but four years' participation by the U.S. in World War II put the nation's human and other resources under much more stress. With millions of men drafted for military service, women were needed more than ever in factories and other places of work outside the home. Limited participation in military service roles was opened to women—as "WACs" (members of the Women's Army Corps) and Navy WAVES (Women Accepted for Voluntary Enlisted Service). Thus women in America faced expanding occupational horizons and were no longer so confined to domestic tasks.

After the war, there was uncertainty about the appropriate roles for women. Contradictions and confusion prevailed. Some occupations seemed to require 'male' strength and aggressiveness. Women continued to be excluded for decades from serving alongside men as firefighters or police officers. Women generally did not get hired as truck drivers or construction workers.

Not until the 1970s, when social and economic changes reinforced a growing women's movement, were there opportunities for women in such occupations. Eventually almost half of the labor force were women. At the same time, male participation in household duties and child care began increasing. The military abolished formerly all-female service *branches*, and the armed forces became more unequivocally gender-integrated in the mid-1970s. Specific positions open to women were limited at first, but by 1990 women were admitted to nearly all noncombat positions and 11 percent of members of the U.S. armed forces were female. In 1991, Congress made it legal for women in the Air Force to fly even in combat.[117] By the time the United. States involved itself in wars in the Middle East, even the news media shifted (along with politicians) to referring almost always to "our brave men *and women*" in combat.

The changed relation of women to military organizations occurred in other countries, too. When Israel had to fight to confirm the independence it had been accorded by the United Nations in 1948, women were crucial. "Over the years photographs of young Israeli women in fatigues carrying Uzi machineguns have reminded us of the legacy of women fighters that began in the first days of Jewish immigration to Palestine. Over twelve thousand women are said to have fought during the War of Independence." Subsequently women continued being drafted into the Israeli army and trained with weapons but were not generally allowed into battle.[118]

In Canada, too, the armed forces moved toward gender integration after World War II. The Canadian Human Rights Tribunal, which hears cases of discrimination against federal employees, ruled in 1989 that in Canada's armed forces all combat positions had to be opened to women, and no restrictions were to be placed on their numbers. Qualified women could become fighter pilots or serve in tanks, and female naval officers could legally serve everywhere except in submarines where mixed-gender crews were precluded by privacy needs.

Like World War II, the Vietnam War gave impetus to gender integration. There were thousands of American women in military service in Vietnam in the 1960s and early 1970s. To a significant extent, however, the roles of military women in that war reflected persistence of the traditional belief that war is "men's business." Throughout the Vietnam War years, several branches of the services recruited at nursing schools. Nurses were the predominant female military occupational category. They were all commissioned officers, some did serve in a combat zone, and they were generally limited to a one-year tour of duty. Navy nurses lived and worked either on hospital ships that sailed off the coast of Vietnam or at bases in Cam Ranh and Danang. Air Force nurses cared for wounded men in evacuation hospitals at those locations and on evacuation flights both within Vietnam and out to Japan, Okinawa, the Philippines, or the United States.

In addition to nurses, there were some military women in Vietnam working as teachers, secretaries and clerks at various bases. There were also female air traffic controllers. Other women were photographers or cartographers. But mostly, the women in military service in Vietnam were either taking care of wounded people, sick people, children, or 'doing' for men—reflecting some persistence of traditional gender expectations.

Obsolescent Pressure to Reproduce

Apart from wars, depressions, technological change, there were other factors involved in reshaping gender-role expectations. One was the enormous increase in population, exerting pressure *opposing* excessive reproduction. The world had more than three times as many people when the 20th century ended as there were when the 20th century began—so in many countries it had begun to be apparent that "be fruitful and multiply" was no longer appropriate as a prime imperative for human beings.[119]

Obsolescence of that former imperative became a major characteristic of recent times. This new circumstance made for pressure to dissociate ever further our proliferating non-reproductive roles from our diminishingly focal reproductive roles.[120]

Whether or not we now like to admit it, the fact that mammal babies are gestated within wombs and nurtured primarily if not entirely (at least for the early portions of their growing up) by the adult female parent had meant in a pre-industrial world that occupational roles for women had to be compatible with their reproductive role as mammalian females. Because of the more extended involvement of a female with reproduction (compared to a male's involvement), as the human need for (and tolerance of) abundant reproduction was diminished by conspicuous population growth, then whatever societal pressures were exerted by evolving cultures toward dissociating occupational roles from reproductive roles would affect women even more than men. Today, in short, women's other roles no longer need to relate so closely to their actual or potential condition of maternity.[121]

One result of these changes and this on-going dissociation of roles was a blurring of the institution of monogamous marriage. As a further result many studies of what has been called "Household Division of Labor" were made. These have dealt mainly with the apportionment of various routine household tasks between husband and wife (or between unmarried male and female cohabitants). In as much as the concern in many of the people doing these studies has been the "fairness" (or "unfairness") of this apportionment, the usual contention (or implication) has been that women bear a larger burden of work (albeit non-reproductive) within the family home. Research findings usually confirm this perception and it is quite naturally regarded as especially unfair in the "working wife" case, where the female partner holds a job outside the home. Even when her outside working hours are comparable to those of the male partner, she assumes or is assigned a disproportionate fraction of homemaking task time and effort.

Valuable as these studies may be in other respects, they appear unconcerned with possible advantages of specialization. Given the particular interests which caused this kind of research to be undertaken, that may be an altogether justifiable change of perspective. But it is nevertheless important for our purposes in this book to emphasize the fact that instead of considering potential efficiencies in assigning different *kinds* of household tasks to each partner, based on possible differences in personal skills and/or interests, the question usually addressed in such literature is whether the two partners share each task equally (or at least, in some sense, equitably). Studies have sought to document and measure a "gender gap."

Contexts Differ

To assess the significance of the change of meaning of the phrase "division of labor"—from Durkheim's usage referring to differentiated and thus interdependent specialists, to feminists' and family sociologists' recent usage referring instead to dedifferentiated husband-wife time-sharing of each common task—we must consider the contrasting contexts. Housekeeping labor, whether fairly shared or not, occurs within what is basically a primary group, the family. Its theme of intimacy makes it fundamentally different from the mercenary context of paid labor. As we shall see in subsequent chapters, the mercenary context of paid labor is a major part of the reason why the anticipated blessings of "organic solidarity" have not been the predominant outcome of industrial level division of labor.

Whether or not studies of household role allocation (often termed "household division of labor") cite Durkheim, what they refer to as "division of labor" is clearly *not* something considered in the sense in which he used the phrase. Some may be tempted to call this more Tom Sawyer-like view that appears in these studies "androgynous division of labor." But that phrase would still beg the question: does it signify *division* of labor into diverse specialties, or does it refer to *sharing* of undifferentiated labor? For Durkheim, the original phrase, in use before serious thought was given to androgyny, unambiguously involved specialization, with workers so differentiated in the kinds of tasks they performed that no one produced the whole marketable product, much less the entire array of goods or services involved in a society's way of life. Thus no worker was self-sufficient. Specialization increased efficiency but sacrificed independence.[122]

As Durkheim viewed the matter, it fostered *inter*dependence. For him that was the point. But for those studying so-called household division of labor, interdependence ceased to be the prime concern, the question of efficiency was hardly in the picture, and fair sharing or unfair imbalance had become the issue.

That recent change of concern shifted attention away from the question of whether division of labor in the original sense still tends to foster social cohesion (organic solidarity), or ever really did. The distraction from that issue was an expectable product of reduction of sex-role differences in paid employment, with fewer job types being exclusively male or exclusively female than formerly.

But now in the chapters that follow we must turn to considering whether modern division of labor (in the original specialization-and-exchange sense) may be pushing us in a seriously anti-social direction?

CHAPTER 6

How We Created New Perils

"It was the best of times, it was the worst of times."
—Charles Dickens, *A Tale of Two Cities* (1859, bk. I, ch. 1)

"By *vice* we mean practices that harm one's self; by *sin* we mean conduct that harms another."

—E. A. Ross, *Sin and Society* (1907, p. 90)

Setting aside the conceptual distraction reviewed in Chapter 5, return now to consideration of the contrasting views of Durkheim and Ross about the societal effects of industrial-level division of labor. Interdependence leads to "organic solidarity," said Durkheim. Interdependence puts us at one another's mercy, Ross argued.

It was in a small but earnest treatise published in 1907, which he titled *Sin and Society,* that Ross voiced emphatic concern about where twentieth century societies were headed. The vulnerability implicit in the interdependence resulting from division of labor leads, he explained, to a multitude of new forms of wrongdoing. For that reason Ross said we needed "an annual supplement to the Decalogue." Old taboos no longer suffice.[123]

Ross knew there were writers who contended *people* were becoming better and there were others who believed human *society* faced worsening troubles. Both might be right, he suggested.[124] Such a paradox was due, he said, to the fact that while human beings could be improving in their personal relations (in what would later be termed "primary groups"), the control of collective activities such as industry and business was becoming ever more impersonal.

As the American nation grew more populous, inventive members of that growing population ensured that its expansion across the continent was accompanied by marvelous technological breakthroughs. Together, the growth in numbers and the advance of technology made for not only commendable individual and collective accomplishments but also new forms of dependence and new ways in which people's daily routines could harm others, often unintentionally but sometimes deliberately. New opportunities for new people in new places with new devices and new occupations produced new temptations and new collateral damage.

Indirect Vandalism

Back in the time when American railroads still powered their trains with coal-burning steam locomotives, my parents happened for a time to be living just a short city block from a railroad's main line. When I joined them there for a year, after my navy service in World War II, it was fun hearing the steam whistle as trains left the depot a few blocks away and gathered speed as they headed out of town through our neighborhood. But I noticed how much more often my mother had to mop soot deposits off the floors and stairs in our house, and dust the tables and chairs and window sills here, compared to previous years when we resided elsewhere. Neither the stockholders nor the executives of the railroad company, nor any of the train crews or yard hands, were vandals who deliberately dirtied innocent people's homes. Yet homes near the tracks *were* dirtied by that era's transportation system. The harm done was the sort of shrugged-off consequence the pharmaceutical industry would later call a "side effect," the military would call "collateral damage," and economists call an "externality." No railroad company provided free house-cleaning service to neighbors along its routes.

From time to time, too, in the early days of railroading, sparks from locomotive smokestacks set fires to farm crops, forests, or buildings near the tracks, and the railroad corporations didn't always pay compensation. Even this kind of damage may not have been considered an egregious "sin" committed by the railroads, but public animosity was aroused by other aspects of the railroad business.

For all the harm inflicted, railroading was a business of genuine accomplishments. Among other benefits it gave the nation (and the world) a better way of coordinating activities in different locations. In 1883 the American Railway Association divided the country into four time zones—shifting, as someone said, from "God's time" to "Vanderbilt's time." Previously there had been twenty-seven incremental measures of time across just the state of Illinois, while Wisconsin had had thirty-eight different times. Standardizing the clocks was essential to building a continent-wide country of interacting citizens, and the railroads got it done.

But railroading was also a business primarily concerned with, as we would say today, "the bottom line"—extracting money from customers, while building its own industrial empire. The obvious benefits to society this rapidly expanding transport system provided by linking widely separated locations to one another were coupled with some major inequities and injustices. There were price-fixing schemes among the various railroad corporations, secret rebates to favored (high-volume, profit-yielding) shippers, and exorbitant freight rates on short routes where a particular railroad company had a monopoly.[125]

Before writing *Sin and Society*, E. A. Ross had had a personal encounter with the power (and the anti-social potential) of railroad industry leaders. From 1893 to 1900 he taught economics at Stanford University. He was fired from that job in a controversy over academic freedom. Speeches he had given supporting the Democratic Party's presidential candidate, William Jennings Bryan, in 1896 had outraged Jane Lathrop Stanford, who was an ardent Republican and the widow of Leland Stanford, former California governor and senator and one of the founder-financiers of the Central Pacific Railroad. Senator and Mrs. Stanford had co-founded the university to honor their late son, Leland Stanford, Jr. Mrs. Stanford demanded Ross's dismissal from the university faculty. The president of the university initially opposed her demand but eventually acceded to it, and Ross was forced to resign in 1900.[126] After teaching at the University of Nebraska for several years, he joined the University of Wisconsin economics department in 1906, and he taught sociology there until retiring in 1937, three decades after publishing *Sin and Society*, which he followed by producing many articles, and several other influential books on a variety of sociological topics.

As modern business enterprises grew, they became more and more impersonal. Being impersonal, they could become irresponsible. There was nothing like distance, Ross explained, to "disinfect dividends." The moral character of a corporation's stockholders therefore mattered very little in shaping the way the corporation's affairs were conducted. A corporation feels no anger, nor can a corporation hold a grudge. If the policies and actions of a corporation ruin someone's life, depriving him or her of livelihood or damaging his or her surroundings, this happens not from any malice, but simply because such a victim happens to stand in the way as the corporation provides its specialized product or its particular service in pursuit of its goal of monetary profit.[127]

By means of what Ross referred to as "the magic of limited liability," corporate business was every year more and more distanced from owners, the stockholders. The expression "absentee owners" had crept into the language (often used with a derogatory connotation) and was sometimes used to signify this separation. Every year, said Ross, these share-holding owners become more numerous and more scattered. Whatever their personal concerns and inhibitions might be, it is management (the "big insiders," as Ross called them) who set a company's actual policies. If savings banks, stock brokers, and insurance companies had increasingly become the impersonal link between corporate management and the millions whose monetary investment gave the company existence, this was a dissipation of personal and moral responsibility.[128] Such indirect connection made it ever less likely for the consciences of any company's "owners" to reach the company's managers and humanize their business decisions.

The Nature of "Sin"

When the actions of any of us are harmful to another, those actions fit the essential meaning of the old-fashioned word sin. Ross recognized the fact that what constitutes sin changes as societies change. Societies are associations of people in mutual interdependence. The practice of mutualism has always enabled sinning, Ross believed. "Most sin is preying, and every new social relation begets its cannibalism." Not until there was buying and selling was anyone able to "falsify the balances." Not until there was lending, could there be usurious interest charges, loan defaults, and foreclosures. Until there was a wage system there could be no wage cuts or laid-off workers. These bitter human experiences occur in that "mercenary context of paid labor" referred to in the previous chapter. Acts called treason may be dysfunctional by-products of the rise of political states but so are counterfeiting of currency, smuggling, and misappropriation of public funds.

"Commerce tempts the pirate, the forger, and the embezzler," said Ross. "Every new fiduciary relation is a fresh opportunity for breach of trust." The sins characteristic of his time (and ours) tend to be "incidental to the ruthless pursuit of private ends" so the "wrongs that infest our articulated society" are essentially acts of betrayal rather than aggression.[129]

Every technological advance opens new opportunities for reprehensible misuse. Were he living today there can be little doubt that Ross would deplore the anti-social side-effects of some of our modern conveniences, and would especially deplore the tendency of merchants to promote use of new apparatus or products with complete disregard for potential hazards or likely misuse. Consider today's personal computers and the internet, more than half a century after the end of Ross's writing career. There is, for example, an obituary Website, *Legacy.com*, on which people post notices of others' deaths. This web site makes it possible to mourn online, but it has had to employ a corps of specialized workers whose job is to prevent people filing unflattering thoughts about the recently deceased. In a room full of computer terminals amid hushed conversations, forty-four screeners study some 18,000 death notes that come in daily. These screeners' task is to reject any backhanded compliments or mean spirited innuendo. Revelation of dark family secrets is taboo. These screeners call it "Dissing the dead"—a temptation that became a costly problem for such web sites where few had foreseen how nasty some of the postings to guest books would be.[130] The internet would have amazed Ross, but the *necessity for such screening* would not have surprised him.

"Modern sin," he wrote, "takes its character from the mutualism of our time." The term *mutualism* seems to have meant for Ross something very similar to what it means for ecologists—a relationship between two or more populations which interact in such a way that organisms of both populations benefit from the interaction.[131] In ecology, the two or more populations would usually be of different species, and their mutualistic interactions could be brief episodes, such as the encounters between flowering plants and whatever insects take their nectar and spread their pollen, or they can be long-term interactions that involve close physical contact with on-going biochemical exchange (e.g. such as the interactions between trees and mycorrhizal fungi). As a specific example consider ant-aphid mutualism: ants may cultivate aphids for their secretions of honeydew (which the ants use as food), while in turn the aphids are protected by the ants against predators.

In a modern industrial society, we cannot avoid entrusting our vital interests to others. In a society with elaborate division of labor we rely upon others to look after our sewage, invest our savings, nurse our sick, and teach our children. As Ross noted,[132] municipal water mains serve in lieu of our ancestors' wells. Public transit takes the place of the carriages of our forebears. A safety deposit box at the bank better serves the function once performed by an old stocking under the bed. Individuals' eyes and noses, and individual citizens' judgment, said Ross, now must defer to the paid inspectors of food, of drugs, factories, tenements, or insurance companies. The "policeman's billy [has taken the place of] my fist," he contended. Today, though, he might have noted other fist-substitutes such as the heightened insistence on "Second Amendment rights," the widespread domestic ownership of hand guns by private citizens supposing their possession provides safety, and the exploitation by such interest groups as the National Rifle Association of public anxiety about crime, together with the compulsion political office-seekers in America often feel to join that organization as a means of averting the stigma of being "soft on crime." In the modern world "sinning" gets complicated and has ambiguities.

Modern Piety: Let us Prey

Interdependence-induced vulnerability is frequently evident in our time in news articles. For example, a strike called early on December 20, 2005 by subway and bus workers had a "severe to devastating" effect on businesses, said New York's Mayor Michael R. Bloomberg.[133] A noon-time drive past storefronts with locked security gates, or a stroll through shops amazingly quiet just a few days before Christmas showed that widespread economic pain was being felt. The mayor denounced the Transport Workers Union, and demanded an end to the walkout. The cost to the city's economy inflicted by the strike was estimated by the city comptroller at $400 million the first day, and at $300 million for each subsequent weekday.

Strikes by labor unions may be situationally necessary means of pursuing legitimate goals. Nevertheless, it is the web of interdependence characteristic of a modern society that presents many downright sinister opportunities, and these have often been seized with little hesitation.[134] In Ross's time "such treasons, "as he called several kinds of exploitative activities, had not yet become infamous. "The man who picks pockets with a railway rebate, murders with an adulterant instead of a bludgeon, burglarizes with a 'rake-off' instead of a jimmy, cheats with a company prospectus instead of a deck of cards, or scuttles his town instead of his ship, does not feel on his brow the brand of a malefactor." Thefts and homicides lurk in the complexities of our social relations but they don't always occur in dark alleys. They may not even involve any overt violence.

When corporate executives were described as high-power dealers of woe,[135] each of whom could sin "with a calm countenance and a serene soul, leagues or months from the evil he causes," Ross's words could almost have been a commentary (a century ahead of time) upon the persistent denial of global climate change by spokespersons for fossil fuel (and fueled) industries. Their refusal to comprehend future planetary damage, and their insistence upon doing business as usual despite innumerable warnings from scientists, would provide a glaring example to include if one were updating Ross's little book. Upon the gentlemanly status of investors, CEOs, and corporate managers "the eventual blood and tears do not obtrude themselves," said Ross. Such persons remain both self-respecting and often even socially respected. No matter how ultimately harmful someone's actions may turn out to be, many things done in the name of, and for the sake of, "business" are not only condoned by leaders of corporations and their stockholders (and by members of government), but are also accepted as normal and inevitable by the general public.[136]

Ross perceptively acknowledged that many modern sins simply consist of *augmenting risk*[137] so heart-rending certainty may not be associated with the "sinful" acts of modern malefactors. When catastrophe happens, the conscience of the modern sinner whose actions enhanced its likelihood is eased "by blasphemously calling it an 'accident' or an 'act of God.'" Looking ahead, that blasphemy will surely be a recurrent misdefinition of our situation when the world confronts enormous woes following our pell-mell depletion of this planet's carbon-sequestering "fossil fuel" deposits and our centuries-long practice of injecting greenhouse gases into the global atmosphere, bringing about the devastation of coastal lands by rising sea level.

People tend to suppose individuals who are recognizably evil in character are the most dangerous, but Ross suggested this would be true only if everyone had equal opportunities to do harm. Were that the case the most villainous individuals would be seen as the worst threats to society. But actually a recognized "ruffian" has limited opportunities to do great harm. Insofar as people do not depend on him, he can hardly commit a breach of trust. Neither the safety of the public, nor the soundness of its money, nor the people's general welfare is in his hands.[138]

Undermining Society Itself

Sometimes the harm done is not to individuals but to institutions. The very foundations of our civilization may be undermined nefariously or perhaps even unwittingly in the ordinary course of pursuing personal advancement or company profit. Modernization has led to denser populations than existed in earlier times. These are able to live in peace only insofar as there is what Ross called "a protecting social order." Anyone who undermines this social order does worse than hurt individual victims. Wounding society itself, he insisted, is a worse sin than directly harming particular individuals.

The complexity of today's societies makes them vulnerable in new ways. There are more ways they can be harmed by the self-centered and short-sighted actions of persons or groups within them. Once the only wrongs against the whole community were treason and sacrilege. People very early developed habits of reacting strongly against these violations. Later on, American frontier communities tended to react promptly with severe sanctions against anyone who provided whiskey to the Indians, or started a prairie fire, or spread diseases. All actions of those sorts were readily recognized at the time as threats to society.[139] Now, though, there are innumerable subtler and more indirect ways in which common interests can be put in jeopardy. Year after year a modern society becomes more vulnerable. And a world becomes more vulnerable when it consists mostly of modern or modernizing societies.

Consider a 20th-century "change for the better" that has brought new problems. The Federal Highway Act of 1956 established the Interstate System of freeways across the U.S. Much as the railroads had done in the 19th century, this 20th-century construction boom helped open vast expanses of the West, where highways like Interstate 70 changed just about everything by putting on the map distant places that had been mostly untouched.

When Interstate 70 was built through the Colorado Rockies in the 1960's and 70's, the area was largely rural and remote. Many people liked it that way, but the older roads—to be replaced by the modern superhighway—were widely recognized as dangerous. Over subsequent years, as the population grew, delays and frustrations on the new highway began to increase. Even at an alpine altitude traffic jams became common. At 8,800 feet above sea level, Silverthorne used to be little more than a gas station rest stop, with a grocery that got fresh produce only once a week. The new highway's increasing traffic volume led to the surrounding alpine meadows being invaded by factory outlet stores. State officials found themselves having to begin considering a major widening project for Interstate 70. But that project proposal was dividing people in the region over the question of whom the highway is meant to serve. Interstate 70 truly transformed this high country area. The transformation was resulting in serious division among the people impacted by it.[140] Among the residents of this portion of Colorado the burgeoning antagonism mirrors in microcosm the reactions to planet-changing impacts of technological progress coupled with modern division of labor.

New Technology, New Specialties, New Harms

Elaboration of the division of labor and advancement of technological culture go hand in hand—technological inventions breed new occupational specialties, and vice versa. An unwanted (and too often unanticipated, and insufficiently recognized) side effect of this technological-occupational "progress" is the escalating societal vulnerability not adequately foreseen by Durkheim but emphasized by Ross. With new ways in which societies are vulnerable, many human activities deemed "normal" within a given culture can inadvertently trigger societal traumas, small or large. But in our time there seem also to be increasing threats from people with overt desire to make *deliberately* destructive use of our increased vulnerability. "Terrorists," we call such people. Occasionally an event that was the inadvertent result of routine actions or common natural events can bring the vulnerability to our attention, and serve as "warning" of what the "bad guys" conceivably might do.

For example, a 2003 electrical blackout in the northeastern U.S. and adjacent areas of Canada was wholly accidental but revealed the vulnerability of North America's electrical grid. Reducing such vulnerability suddenly seemed a top priority. The system needed to be restructured. Key facilities needed to be reinforced to guard against terrorism. Al Qaeda documents suggested terrorists had already considered attacking the grid. To do so, it was now realized, could cause chaos, produce economic havoc, and even cost lives. *A non-terrorist event had produced a foretaste.*

Terrorists could, in effect, put a powerful modern nation back into the dark ages for weeks. The electrical grid distributes energy across a vast territory on an as-needed basis, to power skyscraper elevators, subway trains, ATMs, and many essential things besides lights. Its high-voltage lines transmit power from generators to electrical substations which are controlled by an enormous computerized switching system. That system uses not-easily-replaceable relays to prevent overloads or other failures from crippling elaborate electrical equipment and transmission facilities throughout the grid. The buildings housing the computers and relays are only lightly protected and located next to substations, perhaps surrounded by barbed-wire fences.[141]

That blackout in 2003 left 50 million North Americans without electric power for a while. The grid's fragility was due to a tangle of factors, including the soaring electricity demand together with aging power plants—*and some occasionally rigid bureaucracies.* To eliminate the vulnerability was going to require huge investments of money and time (as well as political capital).[142] But of course there was much institutional inertia obstructing any such investments, so little remedial action was taken, and vulnerability persisted. Again, a comparatively local mishap was providing a foretaste of possibly global future calamities.

From time to time, in various places other than the American northeast, "weather events" put tens of thousands, sometimes hundreds of thousands, occasionally millions, of power-company customers in the dark—not always for a merely inconvenient few hours or days. These storm-damage episodes are reminders that forces of nature, not just malevolent terrorists, can threaten modern ways of living. Such episodes *should* also remind people who often prefer to suppose otherwise that we have allowed ourselves to become far *too* dependent on continued availability of resources that simply will not always be available. Fossil fuels, for example, have already ceased to be cheap, will soon be less abundant, and eventually will cease to be available.

Cultural Lag

Human societies with their institutional inertia are vulnerable also insofar as they are dependent upon non-social phenomena (such as the ecosystems in which they are embedded, systems that live according to laws of nature and are not always conveniently accommodated to human desires). Human myopia (whether innocent or willful) often prevents recognition of such dependence and vulnerability and thus impedes developing policies and practices that would correct or offset ensuing difficulties.

In populous China, for example, deforestation is a serious problem. One small component of that problem is the practice of harvesting annually some 25million poplar and birch trees to provide the wood for making chopsticks.

To begin reducing that exploitation of already scarce trees, the Chinese government recently imposed a five percent tax on disposable chopsticks. It is the chopstick manufacturers who must pay the new tax on about 45 billion pairs of disposable chopsticks produced each year. This cost was expected to be passed on to consumers. In Beijing cafes, there was prompt objection to the new tax. For some the prospect of chopstick deprivation seemed to spell the end of a 5,000-year-old tradition.[143] People—be they Chinese, American, European, whatever—resist ending long-time traditions.

In tandem with the proliferation of vulnerabilities, new ways have developed of camouflaging—as socially admirable virtues—actions that are really sins against the social order. As Ross pointed out, piety can become a masquerade behind which antisocial practices can be carried on.[144] Some instances of this can be understood in terms of the concept of *cultural lag*, introduced into the sociological literature in 1922 by William F. Ogburn,[145] who took note of the fact that a culture's various parts change at unequal rates, some faster, some slower. Arrangements and practices that were once adjusted to each other thus get out of adjustment, and the maladjustment may continue "for a considerable number of years." This cultural maladjustment problem was what Ross was trying to call attention to when he wrote[146] that "The criminaloid flourishes until the growth of morality overtakes the growth of opportunity to prey." He was hopeful that public opinion, as a regulator of conduct, was indeed gaining ground and could one day take over as the capacities of religion and law to inhibit and control criminal behavior declined.

Perceptive as were his diagnoses of the new and egregious forms of misconduct, Ross's hope that public opinion would develop as a grand antidote can be seen with hindsight to have been naive. Change happens, but change is also resisted. Already back then in the first decade of the twentieth century technological progress was so conspicuous and changes in patterns of living were so much a part of life in America that even though Ross noted the lagging societal response to new forms of sin,[147] he grossly *under*estimated the inertia of society and its institutions. The universal tendency of people and organizations to persist in doing what they have been accustomed to doing and to carry on the ways to which they have grown accustomed can seriously delay necessary change. If Ross was far ahead of his contemporaries in discerning the changed opportunities, temptations, and injustices of his time, some of his language habits must nevertheless seem downright archaic to readers encountering his ideas a century later.

The "moral pace-setters" of his time opposed bad personal habits (e.g., beer and Sunday baseball) but Ross indignantly accused them of acting as if big time money-making was something sacred. Insofar as *the most serious iniquities of the day were connected with money-making*, he was appalled that religious leaders mostly stayed aloof from "the big fight." Monopolists and crooked financiers were unintimidated by men whose thoughts about good and evil remained unchanged from those of their village grandfathers. As if expressing resentment still over his firing by Stanford, Ross alluded to "tainted-money colleges" where one might be taught that "Drink, not Graft, is the nation's bane." In religious bodies, personal correctness was exalted above the social welfare of society.[148] How might Ross have reacted to 21st-century revelations of pedophilia among trusted priests, and especially to Diocesan cover-up by quiet reassignment to other parishes? Or, to an administration expending lives of young men and women in a war to control an oil-rich region of the Earth, while adamantly "protecting innocent human life" by restricting research using stem cells taken from microscopic embryos destined otherwise to be discarded.

The Lag to End All Lags

A whole century later, another writer with quite a different personal and professional background, concerned about threats not only to a particular society but to global civilization, had come frightfully close to concluding that the kind of cultural lag so disturbing to Ross was finally threatening the future of humanity at large. Just as Ross had suggested the need for annual revision of the Ten Commandments to make them applicable in the industrial era, James Lovelock[149] at the start of the twenty-first century declared the urgent need for "a new Sermon on the Mount" (with requisite ecological sophistication) and a "new book like the Bible" (drawing upon up-to-date science) *to proclaim the constraints necessary in our time*—to ensure we humans can continue "living decently with the Earth."

This is today's disastrous cultural lag: Our species had grown both so numerous and portions of it so technologically potent that civilized humanity had become a major cause of extinction of other species and a significant factor in global climate change—before the public even began to recognize that our cultural enthusiasm for demographic and economic growth and ecological dominance were serious planetary hazards. Prodigal use of our finite planet by the present human generation threatens its biosphere and has come to be, in Ross's sense of the word, a *sin*—against posterity. What a monumental sin it will have turned out to be—if the historical pattern of taking over for human use ever more of Earth's places and substances (which has deprived many other species populations of habitat essential to their survival and driven them to extinction) culminates in destruction of habitat essential for our own species!

Having dangerously tampered with the very biosphere, we wonder: How could we have fallen into this predicament? It has happened because of other fateful cultural lags—the unseen inter-relevance of separate bodies of knowledge remaining too long apart. Insights of Durkheim about division of labor appealed to different minds than those who were receptive to ideas about ecosystems. Division of labor between academic specialties obscured "the big picture."

We availed ourselves of the marvelous advantages of extensive division of labor before learning to recognize its dark side. Followers of Durkheim and followers of Ross remained disconnected from one another. And both were unaware of the human significance of biogeochemical developments. As separate categories of specialists, those who investigate ecological processes used language and pursued interests that remained foreign to the language and recognized interests of those who promote and manage industry, and most of those who study industrial and other human relations.

CHAPTER 7

How Could It All Hang Together?

> "The division of labor among specialists narrows the range of tasks of individuals and thereby expands the range of tasks that can be accomplished in the society ... There is no direct contact between most persons performing different functions in complex society."

> —Peter M. Blau, *Inequality and Heterogeneity*, 1977,
> pp. 188, 199

We, the talking descendants of apes, use the wonderful capacity for speech both for occupational and recreational communication. Words and other symbols enable us to become differentiated in the kinds of work we do—and in the ways we play.

Our work roles and our play roles sometimes interconnect. Sometimes we can derive enlightenment from these connections, especially insofar as our language-facilitated thought processes enable reminiscence to become analytic—as I will let mine do in this chapter.

The sociology department in a university where I taught several decades ago had a tradition of annually holding a self-deprecating evening party. These gatherings were fun, and some of us at least appreciated them as a valuable antidote to any temptation to take ourselves and our work too seriously. Following a pot-luck "banquet," various members of the department (including graduate students) made humorous speeches or performed skits to satirize or lampoon the sociological research or classroom activities to which we were more seriously devoted five days each week. On one such occasion a young faculty member, challenged to comment on "whatever passage from sociology literature had most inspired" him, quoted with a straight face the following sentences from a textbook (written by the professor who was then head of the department, and was present in the laughter-anticipating audience)

> This is what we mean when we call man a social animal. He does not have the ability of a salmon, to hatch from an egg and make his own way in life.[150]

Delivered so dead-pan and out of context, these "inspiring" words seemed hilarious (even to their author!). It is nevertheless a fact (which we all recognized in our more routine scholarly moments) that no human, especially in the context of a modern industrial society, is or can be self-sufficient. We all depend in myriad ways upon many others.

It is often far from obvious how dependent we are, nor do we easily recognize the vastness of the range of consequences flowing from our interdependence. It is more than just the fact that there can be strength in numbers; there are advantages to *diversity*.

Retrospectively Discovering Interdependence

Let us therefore explore the scope and subtlety of our interdependence. One summer day during my adolescent years, I was invited to go sailing on a nearby lake with a high school friend and his father, in a boat they had recently built. It was possibly the boat's "maiden voyage." My friend and his dad were justifiably proud of their achievement. As water craft go, however, it was rather primitive, no bigger than a small rowboat, flat bottomed, with a retractable center-board in lieu of a keel, and a small mast and boom supporting a sail just large enough to capture modest amounts of wind power. They were well aware that it could never suffice for navigating any great distance, but it was eminently suitable for short pleasure jaunts on small bodies of water.[151] Memory traces of that adventure have remained in my head through six and a half decades mainly because of a laughable concluding incident that would be unlikely to have occurred were this not the very first time this father-and-son team of builders had sailed their boat.

It was a comic ending that firmly established for me an appreciation of Newton's third law of motion. After some two or three hours of leisurely tacking back and forth to employ the energy of a gentle breeze, we returned to shallow water alongside the dock, with the center-board raised. My friend's father was in the stern at the tiller, I was seated in the middle of the boat, and my friend was standing in the bow holding a rope with which to tie up the boat. He prepared to step onto the dock—across a very narrow band of shallow water. What would have been a short, small step even for a teenager became to his astonishment appreciably longer because he forgot those words of the perceptive Isaac Newton we had been taught in physics class—that for every action there is an equal and opposite reaction. The boat was small enough and so light that as he casually thrust his right foot toward a spot on the dock, his left foot pushed the boat away—just enough so that instead of ending up at the expected destination on the wooden platform he found himself standing nearly waist deep in water between boat and dock! What made it so memorably funny was his "body language"—motions so casual and relaxed it had looked as if stepping right into the edge of the lake was just what he meant to do!

It was natural to admire the fact that the two of them had built the boat themselves. Modest vessel though it was, however, its existence owed much to contributions from numerous other unseen persons. Neither father nor son had been involved in publishing the magazine containing the plans for producing a home-constructed sailboat. They had not personally harvested the trees from which workers in a sawmill had created the boards they bought from a local lumber yard. Nor had they mined the iron ore that was somewhere smelted and elsewhere combined with other ingredients to make the steel which, in some factory, was turned by others into the various tools the two of them had used to assemble some of the lumber into the simple hull, and convert other sturdy pieces of straight-grained wood into mast and boom. Nor had they woven the canvas from which the sail was fashioned, nor twisted the hemp into ropes for rigging the mast and controlling the sail. Their home-made boat was a product of many skills other than their own.

Thus, implicit in my memorable adventure there was more *division of labor* than my eyes beheld during the brief display of amateur "seamanship" on that local lake, and much more even than I would have seen directly had I been present in their home shop to watch the father and son assembling the boat. Only upon thoughtful contemplation many years later would I recognize that some of the extensive division of labor so taken for granted in a modern society had been reflected in the materials, tools, and design employed in this "do-it yourself" project.

Their efforts had brought forth a pleasure vessel quite respectable but of strictly limited utility. However enjoyably it could sail the waters of a small land-locked lake—during which its builders might *dream* of oceanic adventures—never could that little boat fulfill such dreams. It was too small to be sufficiently seaworthy even to venture out onto "the big lake" (nearby Lake Michigan), much less for attempting any transoceanic voyaging. We three were all familiar with the comings and goings of large ore-carrying boats[152] on the Great Lakes, and vessels of that size frequently came into the local harbor, past one or more upraised drawbridges, and up the river that ran through our town, either bringing loads of pulpwood for the paper mill, or coal for use in the town's various furnaces, or to receive a load of the clean sand that was then available in abundance from the dunes (now largely gone) along that stretch of Lake Michigan's shore. For several decades those dunes provided sand for foundries elsewhere in the Great Lakes region, though it was such clean light hued sand we also supposed some of it was turned into glass that would become windows in some of the nation's homes and automobiles.

Aspirations to venture farther away over water were for me later fulfilled when, with a cousin and his family I enjoyed a crossing of Lake Michigan aboard a coal-burning car ferry operated by and for a regional railroad. While it happened to be the smallest of that company's fleet of five car ferries, its cargo deck had four parallel standard gauge railroad tracks, onto which were loaded (and securely fastened in place) some two dozen freight cars. It also carried a few automobiles, with passengers accommodated on an upper deck. An appreciably greater assortment of eminently watchable skills were exemplified by the crew of this steel steamship than were displayed by the two amateur seamen sailing that little wooden boat, and part of the pleasure of the journey was witnessing prior to departure the loading operation, replete with considerable evidence of meticulous coordination of the diverse actions of differentiated specialist workers employed by the railroad.

Lake Michigan is wide enough so that both shores are actually out of sight briefly at mid-crossing, when fantasies of oceanic voyaging are briefly nurtured. Today, more than in those car ferry years, some of the shipping on the Great Lakes does actually come from overseas. Use of the St. Lawrence Seaway has caused shores of the Great Lakes to become commercially the "north coast" of the continental United States.

Devices adequate for truly global navigation have became common in recent centuries, both for movement on (or under) the ocean surface—and (within my lifetime!) high above it, on wings of aluminum. The international traffic so taken for granted by now represents all the more emphatically the results of intricate webs of interconnected specialized occupational roles far beyond what one sees in using (or even in observing the manufacture of) the elaborate artifacts by which the present era's commercial transport of people and goods takes place. *Globalization* has entered the vocabularies of both its proponents and its adversaries. Division of labor as a fact of human life has been elaborated vastly further than could have been imagined when scholars first began to study its causes and consequences. Today it is international.[153]

Connecting Two Hemispheres

Its enormous influence in shaping modern life is poignantly evident in the fact that my wife and I were recently able to attend two happy events *occurring approximately six months apart—both in early summertime!* The experience involved two summers separated by only *half* a year. Such timing seems contradictory but isn't, because the two events happened on opposite sides of the equator. One occasion was the graduation of our eldest grandson at a university in England. The other was the outdoor wedding of our eldest granddaughter (his sister) in New Zealand. The university graduation took place in June, at the onset of summer in the northern hemisphere, while the wedding occurred in the following January, the equivalent season in the southern hemisphere.

Both events were witnessed by *international* assortments of the participants' friends and relations who rather easily gathered together from all around the planet. In pre-aviation centuries, few people could have traveled to opposite sides of the planet for occasions so closely spaced in time. Our extended family happens to have connections to both New Zealand and Great Britain, but humans are neither born with wings nor can they swim so far. Granted there were remarkable people who traveled by outrigger canoe to populate islands scattered across the Pacific in the vast Polynesian Triangle,[154] but not until modern times were two hemispheres made so accessible to each other as they became in our lifetime. However, we moderns who fly to distant locales should remind ourselves that we do so only with the "assistance" of innumerable other people collectively providing "prosthetic" wings for our occasional use. These two happy in-gatherings, spatially distant but temporally not so far apart, thus reflected the extensive interdependence so characteristic of life as it is known today.

That interdependence is a product of the remarkably elaborate division of labor that developed in the twentieth century. Many, many people in almost unimaginably diverse lines of work, when their differentiated skills come into play in some coordinated fashion, make it possible for each of us to have experiences (or things) that were impossible in earlier, simpler times. No one is self-sufficient. Users of modern apparatus and services neither need to invent them, nor need we be able to provide them by our own efforts. Marvels of modern life happen because there is (a) extensive *division of labor*, and (b) *exchange*. As we shall see, the exchange side of this arrangement is fraught with grave peril.

No individual among my extended family happens to be engaged in any occupation pertaining to the commercial aviation industry. None of us has personally worked for the major airplane maker Boeing, nor the jet engine manufacturer General Electric, nor Alcoa the producer of metallic aluminum, nor any airline or travel agency. We've never worked for the petroleum industry, without which the planes would have no power to fly. All of the friends and relations who gathered for those two events in those two antipodal places, arriving by means of commercial aviation, were linked to that industry "merely" as some of its occasional customers. More will be said later about the manifold implications of "customer" status in the intricate system of interdependence we call society today (see especially Chapters Eight and Twelve).

Appreciating the full relevance of modern division of labor to this sequence of experiences, though, involves not only recognizing how today's swift passenger travel by air involves the coordinated actions of thousands of people, doing hundreds of different tasks. In this instance it also happens to involve knowing why those two countries, England and New Zealand, geographically as far apart as two nations can be on this planet, retain historical connections providing a context for two such events to have happened where they did. Without the political, cultural and language linkage of the two countries it is rather improbable that one family member would have attended university in England, and his sibling residing in New Zealand would thus have married there on the opposite side of our planet. Understanding the connections depends on acquaintance with some facts from the history of the British Empire. As it happens, the connections between these two countries clearly reflect the interdependence that is characteristic of a diversified labor force. In them, we see modern division of labor in operation.

History, Technology, and Interdependence

After the existence of those islands between the Tasman Sea and the southern Pacific Ocean became known to Europeans and were given the name "New Zealand" by a Dutch explorer, it was chiefly British people who settled there. They transplanted the English language, British culture, and European ways of making a living to that distant colony. The British Isles (and Europe) were already "fully settled" and the spread of English people to new land as far away as the antipodes was merely a continuation of the dispersal of *Homo sapiens* that had begun many millennia earlier in Africa.[155] Coming from Britain by sailing ships they gradually outnumbered the prior Polynesian inhabitants, the Maori people.

With the passage of time, distance *could* have weakened these British emigrants' bonds with Britain, especially when there was no quick communication between locations halfway around the world from each other. As an example of the travel-time measure of distance in previous centuries, consider how seriously remote England was from New Zealand—on quite the opposite side of the globe. Charles Darwin, as a young English naturalist learning important things about the world's geology and biology, made a nine-day visit to a small portion of New Zealand's North Island in December 1835, near the close of his sea-going years aboard the 90 foot long two-masted brig *H.M.S. Beagle*. That ship took 278 more days after that short stop-over in New Zealand to get him back to England. This elapsed time of over nine months included 80 days of further stopping in various ports along the ship's less-than-direct route, but it was under way at sea for 198 days altogether. That's one measure of the size of this planet and the remoteness of N.Z. from the U.K. Just to cross the Tasman Sea and reach Sydney, Australia, their first stop after leaving New Zealand, the *Beagle* sailed for 13 days.[156]

Without support from the on-going exchange relations characteristic of commercial interdependence, cultural commonality between far away colonials and the land they continually alluded to (albeit decreasingly in recent years) as "home" might have waned far more than it has.[157] However, before the various countries of the world were linked by cable or radio, and when sea voyages—in transition from sail to steam—were still slow and arduous, certain events in 1882 established what ought even now to be seen as a remarkable fact: populations at maximum earthly remoteness from each other *could* create, retain, and expand economic interdependence. In February of 1882, the *Dunedin*, a refrigerated vessel, departed from a Port Chalmers wharf in the Otago Harbour on New Zealand's South Island, beginning a *three months* voyage to London where it offloaded a cargo that included 4,460 frozen mutton carcasses and 449 frozen lamb carcasses. The meat is said to have arrived in perfect condition. Despite the quarter-of-a-year transit time, refrigeration thus enabled New Zealand farmers thenceforth to sell meat (not just wool) in markets on the opposite side of the world.[158] So farming down under became considerably more profitable, and people in Britain could consume more meat than there was land available at home to produce.

Opening of a new market for the flesh of New Zealand-grown sheep had other consequences besides augmenting the animal protein portion of British diets and filling the wallets of New Zealand farmers. New Zealand farmers became more differentiated. Those holding the best land were nudged away from raising Merino sheep for their wool, and towards running other breeds such as Romney, to provide both meat and wool. Merinos were increasingly relegated to high country sheep stations. The ships that carried meat and wool to the British market brought various manufactured goods back to the New Zealand market. Often, as "ballast" they carried loads of iron roofing for the homes New Zealanders built.

Technological progress was important. Refrigeration technology launched a major change in global exchange. As time passed, "experience enabled the charges for freezing, shipping, and marketing to be lowered. This encouraged the farmers to ship more meat and so helped to bring the charges down still further. Thus the industry gathered strength. In 1882 the value of the frozen meat exported from New Zealand had been £19,000; by 1890 it had increased to over a million pounds, and by 1895 to over a million and a quarter." [159]

In New Zealand, the need to slaughter and prepare animals for export in greater abundance fostered increments in the division of labor. It led to creation of that country's first large-scale industrial plants, the freezing works. As development of railways shortened travel times within New Zealand, wider domestic distribution of products also induced growth of various industries, leading to new occupational specialties. Some new industrial plants processed such by-products as sheepskins, tallow and manure. Several of the country's smaller seaport centers—Bluff, Oamaru, and Timaru in the South Island, and Gisborne and Napier in the North Island—saw increased traffic from the enlarged sheep industry. As the division of labor within the small country down under thus escalated, the interdependence of people *within* that land as well as between them and people overseas grew substantially, and fostered growth of urban centers.

With refrigeration it was also possible for New Zealanders to export butter and cheese, formerly items they produced solely for their own consumption. In the early decades of the 20th century, the quantities sent abroad rose impressively. Most of the butter and cheese exports, like the meat exports, went half-way around the planet to Britain, making New Zealand more and more the "dairy farm of the Empire." [160] All of this depended upon, and contributed to, further ramified division of labor, which led to New Zealand becoming one of the world's most urbanized countries[161] even while its principal exports came from farms. The need of diverse specialties to interact with one another gives rise, in whatever country, to urban concentrations of population.[162] With the growth of urban concentrations of people came various otherwise unsupportable amenities, such as concert halls or opera houses, major art museums, book publishers, universities, etc.

So we see another of the remarkable things about *Homo sapiens* as a species. With advancing technology we have developed a nutritional system that enables enormous numbers of us to obtain sustenance routinely from faraway places. In *any* substantially urbanized country, no matter how little thought its citizens today give such matters as they periodically buy groceries at local supermarkets, their lives depend on sustenance shipped in from places far beyond their horizon. (Many Americans were astonished to learn from "health warnings" in 2007 that a number of familiar food items they were routinely buying actually came from overseas—including seafoods and produce from China!) We differ from our chimpanzee cousins in our "foraging" as impressively as in our communicating. Among modern humans, hinterland residents producing the sustenance displayed for us to "gather" from supermarket shelves depend on obtaining other products and services from the cities. *Exchange is endemic in human living.*

Our exchange networks have grown more elaborate in recent centuries. The NZ-UK link is merely a dramatic and extreme case. The globalization that had become so controversial by the close of the 20th century was a measure of the elaboration and institutionalization of exchange relations among many portions of the planet's widely dispersed human population. Truly extensive division of labor had become an essential aspect of these relations.[163]

Flying With Wings of Many Others

Opera-goers realize that people never seen on the stage are essential to the performance. It becomes obvious whenever the spotlight tracks the diva as she moves about; someone has the job of manipulating that spotlight appropriately. But roles usually unseen are even more true of today's commercial aviation. More different people function behind the scenes than we tend to be aware of during a flight. They include mechanics and technicians who conduct preflight maintenance, and those who fuel the aircraft; the check-in staff at the airport (and nowadays, of course, the security staff). Even the aircraft-operating crew consisting of captain, first officer, and perhaps a flight engineer, we don't usually see. We do see the cabin attendants. Mostly unseen are the people who prepare and load the aircraft, both the ramp service team and a cabin service team. We are also unlikely to see the many people involved in communication between cockpit crew and controllers on the ground or many of the airport personnel responsible for a safe landing approach, others responsible for guiding the plane to its assigned gate, and still others who unload luggage and move it to the airport's baggage claim area. We tend only to become concerned about these latter people if—for some reason the airline cannot explain to us—our luggage goes to some destination other than our own!

As airline customers flying to England and later to New Zealand, we had purchased a portion of the services of many different people—not only the skilled persons at the controls of the jet aircraft on which we flew, and the male and female cabin attendants, the ticket agents, the air traffic controllers and radar operators, the jet engine mechanics, electronic technicians, and other maintenance personnel that obviously made our trips possible, but also, indirectly, the services of thousands and thousands of other people in many other specialties. Some were involved in the mining of bauxite, others in the electrolytic smelting of aluminum, or in the manufacture of aircraft parts and the assembly of airliners as well as the myriad airport baggage-handling equipment, and in the elaborate organization of the oil, food, and communications industries upon which airline operations depend utterly. The ape-derived hands, and the evolved and trained minds of all these other people, had enabled us, with our forelimbs (like theirs) retaining unbird-like versatility, to fly great distances buckled into jetliner seats.

Still other people had roles in the scheduling and regulation of air traffic, the legal regulation and facilitation of corporate behavior, in the flow of material products among this array of interconnected activities that add up to commercial aviation. Efforts by these people and more, skilled in one line of symbol manipulation or another, contributed to our being present in those two places at those two times.

All these differentiated occupations without which our easy and convenient travel experience could not have happened, plus the global banking system and the flow of money throughout the vast network of economically linked corporate bodies without which those many occupations would not effectively interconnect, remained quite remote from the thoughts of passengers flying nonstop overnight from Seattle to London, or from Los Angeles to Christchurch. They were even more emphatically absent from our minds as we listened to the Latin incantations at the university graduation ceremony in England or the exchange of loving vows at the outdoor wedding in New Zealand. Upon later contemplation, however, it is clear to be seen—it took modern division of labor to enable all this to happen.

The web of interdependence turns out to be far more complex than at first imagined. Our flight to England landed at London's Heathrow Airport, a facility today's travelers just accept as given, though even a moment's thought would suggest it was as undreamed of in Queen Victoria's nineteenth century as it was thirteen hundred years earlier in the days of legendary King Arthur and his round table. At home after the flight a library source informed me of some interesting facts I had not realized before. Using wartime emergency powers during the Second World War, Britain's Air Ministry acquired some 2,800 acres west of London, including most of the village of Heathrow—and an existing airfield. Those who proceeded soon to convert that area into today's international airport showed remarkable foresight in planning its runways and taxiways so that they can cope with aircraft as large as the Boeing 747, a complex mechanical device scarcely imagined as recently as the 1940s and 1950s.[164] But there have been important changes over the years, at Heathrow and elsewhere. Today at every major airport, information about flight arrivals and departures, with gate numbers, is available on numerous television screens, and passengers mostly move their own luggage about on trolleys (carts) rather than relying on Skycaps (human porters).

It can take thousands of people following dozens of different occupations just to *construct* an airport to serve jet airline traffic, and many others to carry on its myriad operations. All but a few of them remain unseen, unnoticed, or unappreciated during our passage from car park to boarding gate. Shadows of the division of labor are implicit wherever we go. Passengers at major airports everywhere seldom walk on the ground any more between terminal building and aircraft to ascend movable stairs rolled up beside the plane. At Heathrow, as at other international airports, passengers board their flight (and deplane at their destination) through corridors, elevated and on wheels, that connect the waiting area directly to the airplane's door. These telescoping "jetties" (British term) are called "jet ways" in America—a term identified by the Encarta dictionary as a trademark (without specifying who is the manufacturer claiming it). Thus even the simple act of going to the plane from the terminal, or vice versa, is enabled by the labors of other unseen workers. When crossing either the Atlantic or the Pacific while seated on an airplane so huge it feels more like either a ferry boat or perhaps a mobile theater, thoughts may revert to earlier versions of flying machines. It is startling to realize that my own first flight (in a barnstorming Ford Trimotor that came to the Michigan town where I was in high school) occurred almost *two-thirds of the way back to the very beginning* of heavier-than-air aviation. That was less than a quarter century after my father had worn wings of gold as a U.S. naval aviator, piloting blimps at a time when lighter-than-air aviation still seemed a viable rival to heavier-than air. Such reveries may drift even all the way back to the 1903 spruce and ash and linen device with Charlie Taylor's engine turning two chain-driven propellers that carried one Wright brother at a time over the sands at Kill Devil Hills.[165] There is a kind of poetic symmetry in the following pair of facts: (a) those aviation pioneers from the Dayton, Ohio, bicycle shop had to perform carpentry tasks and erect a building on the coastal sands of North Carolina in which to assemble their 650 pound Flyer; and (b) Boeing's development of the 747 jumbo jet had to start with putting together the world's largest factory building. The many building trades, not just the skills of aircraft engineers, mechanics, and flight personnel, were part of the division of labor leading to facilities, devices, and organizations that would enable us to fly nonstop between Seattle and London or between Los Angeles and Auckland or Christchurch.[166]

Although elaborate division of labor and its results were commonly taken for granted by the onset of the 21st century, the diversity of economic roles is so extensive that when one does stop to think about it one begins to wonder how any system of human interaction so enormous and intricate could ever have developed. How could it all hang together? How could members of a single species become so diverse and how could their fantastically diverse occupational activities be so effectively integrated into a functioning system?

Here are facts that are truly fundamental: Biologically we remain a single species, but the processes sociologists refer to as *social differentiation* have wrought remarkable diversity within this one species. Just as patterned interactions among diverse species (food chains between trophic levels, competition, cooperation, predation, mutualism, etc.) enable an ecosystem to thrive, so the patterned interactions among many categories of socially differentiated humans are basic to modern societal existence. The mechanisms that enable the multiplicity of actions by extensively differentiated humans to mesh effectively can be straightforward or subtle. There had to develop a network of *exchange* relationships among these enormously numerous and impressively differentiated specialties. That network of trade would take on a complexity commensurate with the intricacy of the occupational web.

The Unseen Web of Exchange

No worker in any aluminum rolling mill is likely ever to have started his day's labor with the thought, "I'm going to spend the next eight hours helping to make a product that will combine with the products of hundreds of other workers in dozens of other industries to enable the grandparents of two young people to rendezvous with one of them in England and with the other a few months later in another Commonwealth country in another hemisphere." Nor would the engineers designing jet engines have been likely to say any such thing to themselves as they sat down at their drawing boards. Moreover, though perhaps a little less obviously, no such notion would ever be attributed to the conscious but routine thoughts of *any* specialized segment of the labor force, by a social scientist studying phenomena like the division of labor. Each workday for most people is spent just "earning a living." So what connects them? How do their efforts *get* coordinated? A connecting web does exist. But on a day-to-day basis the people participating in this system tend not to be concerned about, or particularly aware of, the nature, the causes, and the many direct and indirect *consequences* of their interconnections.

To this day, and familiar as the sounds and sight of airliners passing overhead have become, and familiar as we all are with the existence of large airports located on the outskirts of major cities in most nations, not everyone flies to opposite sides of the world within a lifetime, let alone within a half year. What explains the inclusion of any of us among the still small minority of this planet's human population for whom such an experience does actually happen? How was it that my wife and I could "afford" those trips to England and New Zealand? (For that matter, what is implied in the very fact that the technical feasibility of such an experience is assumed, so we think about it just in terms of "affording" it?)

Being able to take those flights was due not only to the existence of commercial aviation and all the other attributes of modern taken-for-granted technology and organization, but also to the fact that there had been a "market" for the "ideas" and "knowledge" I had expounded in lectures and publications throughout my years as an academic prior to retirement. In effect it could be said that I exchanged written and spoken *words* for that vast panoply of services listed above that enable a person to fly from point A to point B.

But the exchange, like other exchanges today, was accomplished in a way that disguised (or concealed) the implicit barter—"I'll trade you so much of my product for so much of your product." In pre-European New Zealand, greenstone (jade) was valued by the Maori who made fine chisels out of it, as well as ornaments, and those Maori who lived in areas where greenstone was found traded it for other needed commodities provided by other Maori people in other parts of the islands.[167] For many centuries trading among humans was that simple and straightforward. But barter gradually became more difficult because it became more complicated—as occupational differentiation increased, and as the multiplicity of commodities, products, and services people wished to trade, increased.

So this particular exchange with the airline involved more than just a recipient and a dispenser of tickets. It would be absurd to imagine obtaining our tickets for those flights in a simple person-to-person exchange by presenting copies of my books and papers, or giving any 50-minute lectures, at the ticket counter. An airline ticket agent would be unlikely to have much interest in my kind of "product," certainly not while working to earn his or her living. Nor would the dealer in automobile tires who wished to fly off to a vacation rely simply on offering the travel agent so many tires in exchange, or the surgeon an appendectomy, or the soprano five arias. The days of simple barter as the dominant means of exchange are long gone, absurdly inadequate in the face of changed circumstances.

Even in feudal times, long before human flight was achieved by those brothers from the Dayton bicycle shop—even before it was any more than a dream, and before European cultures had spread to other continents—the system of barter was already becoming increasingly multi-dimensioned. Eventually, as the diversity of goods and services increased, and the diversity of occupations providing them multiplied, the exchange networks became far *too* multi-dimensional for overt bartering to handle.

The need for a generalized *medium* of exchange loomed ever larger. Instead of the farmer trading eggs for shoes at the cobbler's shop, or the cobbler taking shoes to the farm to exchange for eggs, each "sold" his particular product to whomever, took the *money* and "bought" some whatever from whomever else. Coins began to be minted more than two thousand years ago, but only much later did most exchanges by most people come to be done through circulation of money.

The Money World and Its Effects

Money eventually depersonalized exchanges. That is the dark side of our interdependence-fostering division of labor. Subsequent chapters will detail how dark this dark side can be. As money came to be involved in nearly all trading, exchange networks evolved into markets. This facilitated the on-going diversification of the trading network. Especially from the time of the industrial revolution onward, trade became far too complex to be done by any other means than the flow of money. Accordingly, the acquisition and spending of money had to become a central feature of life in human societies. And the once palpable "flow of money" ascended to new heights of abstraction as the ingenuity of financial establishments and the technology of communication burgeoned.

To obtain those airline tickets I had to have sold my product for money. That is, having received a salary or royalties, I could spend some of that "income" for portions of the commodities or services other specialists had produced to "sell." When I "bought" our tickets I was thus exchanging some form of money, rather than my product or my labor, for those certificates of entitlement to seats aboard scheduled flights. Such certificates may themselves be pieces of printed paper, or in the ultramodern case, invisible (i.e., electronic). In whatever form, they serve to communicate to appropriate others at the airport that we have obtained the right to partake of a small but personally adequate portion of those myriad services mentioned in previous pages of this chapter, which, all properly put together and appropriately sequenced, get us to chosen destinations.

In retirement, modern members of industrial societies may continue to receive monetary income, often by "direct deposit" (electronic impulses addressed to a local bank, periodically boosting the number that signifies our "account balance") so—by the manipulation of symbols—we go on exchanging portions of our pre-retirement product for portions of the often vastly different products and services provided by others (not yet retired) in today's labor force. Money, in whatever form, is a category of symbols, and some of the goods and services exchanged by means of money also consist of symbols of various kinds.

Symbols are vital to human societal life, but symbol systems remain mysterious to many, and don't always work as expected or intended. Symbols—words and numbers and various other representations—are quintessentially human phenomena. The interactions of humans are distinguished from the interactions of other creatures by the fact that we make and use various kinds of symbols. To paraphrase and expand the aphorism from Descartes: We speak, we write, we calculate, therefore (or thus) we are human.

Eve Spoke is the astonishing two-word title of one of the many recent fascinating books expounding knowledge about how evolution brought forth the human species. It can be exciting to think about the abundant ways in which our species' mastery of symbols, including words, has so magnificently enlarged our capabilities. At the same time, we need reminders of various ways in which the symbol-using habits of *Homo sapiens* can go awry and become dysfunctional. Simplistic and emotional miscommunication, as portrayed in *The Sound Bite Society: Television and the American Mind*, together with studies of persuasion, multifarious forms of deception, stereotyping, and "disinformation" in modern life, raise important warning flags. The amazing functionality of that network of exchange relationships we call "the market" entails some seriously dysfunctional "side effects." To understand both the functioning and the fact that antisocial side effects have apparently become inevitable, we need to consider explicitly the symbol systems without which networks of exchange relations between differentiated people could never have arisen.

CHAPTER 8

How Human Nature Was Corrupted

"... natural selection ... doesn't stop with its direct productions, such as feathers, ears, and brains. Once natural selection has produced brains ... those brains can go on to produce technology ... [and the] indirect rather than direct productions of natural selection can burgeon into new reaches of complexity and elegance."

—Richard Dawkins, "Intelligent Aliens,"—
in John Brockman (ed.), *Intelligent Thought*,2006, p. 100.

"... our generation ... has ... witnessed ... incredible suffering, brutality and destruction [so we] have been searching for a more realistic interpretation of ... the fundamental nature of man."

—Hobart Edgren, *Of Marble and Mud*, 1959. p. 117

An idea valid at one time may cease to be valid when the world has changed. The time has come for stating a principle very unlike Durkheim's view. His ideas about society and the division of labor seemed to make sense at the end of the nineteenth century. That was a simpler era. Some of its ideas were obsolete at the beginning of the twenty-first century, an era of much more complex experience. Instead of division of labor being the great producer of organic solidarity, it now appears that *division of labor corrupts*. Relations between differently specialized persons in today's societies are increasingly like those between predators and prey.

Probably very few incumbents of specialized occupations today are aware of this cognitively, but many feel estranged by it, without quite knowing in what way the feeling arises (or what they are estranged *from*). Division of labor today corrupts language (humanity's distinguishing attribute), making it less a means of communication to facilitate cooperation and increasingly a means of miscommunication as an instrument for exploitation of others. Division of labor promotes the use of words (and other symbols) not for the purpose of informing but for the purpose of enticing/seducing/exploiting.

Predation Universalized?

To repeat, Durkheim's belief that division of labor could be a source of solidarity was based on his recognition that specialization led to interdependence. But interdependence was seen differently by Ross, who said that insofar as it puts us at one another's mercy it brings about a multitude of new forms of wrongdoing. Ross had a flair for the dramatic phrase, so he suggested that "every new social relation begets its cannibalism."

In more scholarly language half a century later, Richard Emerson noted aspects of power and dependence in systems of interaction.[168] In any system in which there is division of labor, one specialized person's (or group's) power resides implicitly in the differently specialized other person's (or group's) dependency. Actor A is *dependent* on Actor B to the extent that (1) A desires goals mediated by B, and (2) to the extent that A's satisfaction of those desires is unavailable from alternative sources outside A's relation with B. By a remarkably simple formulation, then, Emerson brought penetrating new insights to bear on Durkheim's topic,[169] though his point of departure was not Durkheim's work as such but the sociological understanding of power and authority derived from the work of Max Weber, and Durkheim was not cited in Emerson's seminal paper. But in addition to shedding a different light on division of labor as seen by Durkheim, Emerson's formulation can be regarded as reinforcing Ross's concerns.

If, according to Emerson, the power of A over B is equivalent to the dependence of B on A, then conversely, the power of B over A is equivalent to A's dependence on B. And these power-dependence relations may be either balanced or unbalanced. They are unbalanced when one party is more dependent on the other than vice versa.[170]

Traditionally the phrase *division of labor* has denoted a web of interdependent specialists. In light of Emerson's analysis, it becomes a web of power-dependence relations. And, as I see it, in an industrial society the very intricacy of the system produces a tendency for power-dependence relations to become unbalanced. And recognizing this should begin to clarify who was more nearly right—Durkheim or Ross—about the ultimate effect of an ever ramifying division of labor.

Progress or Entrapment?

Consider for a moment the horticultural revolution in Neolithic times. That was the change in mode of gaining sustenance by which *humans undertook to function as managers* (not mere members) *of ecosystems*. From the present vantage point we can ask: Was it a great leap forward or a commitment to eventual disaster?[171] Likewise we can ask: What about the industrial revolution, when we began in earnest *to harness fossil energy to human tasks*? That revolution was a development that would lead to exponentially advancing technology, vast urban complexes, enormously ramified occupational specialization, intricate (and increasingly global) webs of exchange, interdependence—and mutual exploitation. So, were we thereby opening the gates into earthly paradise or stepping onto a steepening slippery slope?

Durkheim in 1893 wrote so hopefully about the new form of solidarity. But organic solidarity, resulting from interdependence derived from diversity of occupational specialties, could only be "normal," and overall an asset to humanity (i.e. functional), if its main effect were something more than entrapment of human individuals in a system that basically deprives them of any semblance of self-sufficiency. For Durkheim's favorable expectation about the effect of specialization to prevail, the kind of mutually predatory "interdependence" that so appalled his American contemporary, E. A. Ross, would have to be a rare or at least incidental consequence.

Are acts of predation by humans upon their fellow humans rare, or just rarely recognized as such? According to Ross, judgments of ordinary citizens about the conduct of others (apart from close personal relations) tend to be casual, inconsistent, and not really thoughtful. As a result, he said, "the public heeds the little overt offender more than the big covert offender."[172] Acts that "wound society itself," disrupting the social order, worried Ross more than actions which hurt particular individuals. And he argued that there come to be more ways of harming a society as it grows more complex.[173] Were he living today, and familiar with the knowledge amassed since his time by ecologists, climatologists, and geologists, Ross would very likely be furiously writing about the ways modern prodigal living destroys the planetary conditions upon which human societies' continued well-being utterly depends.

Because Ross was using a vocabulary unlike Durkheim's, the issue between their opposing expectations was less than explicitly joined. Social scientists since their time have simply never settled a dispute they did not acknowledge. In later social science publications Ross's polemic has been less often cited than Durkheim's study. But citations of *The Division of Labour in Society* have appeared in contexts not addressed to the issue we are now examining. For example, Durkheim's book was cited in articles pertaining to 19th century working class politics,[174] to economic sociology as a re-emerging field,[175] to the question of whether his work was rooted more in French or in German intellectual antecedents,[176] and even to a comparison of alternative explanations for protests against school bussing.[177] None of these, nor any of the papers in an issue of *Sociological Forum* (March 1994) devoted to a centennial revisit to "sociology's first classic," address the question of whether organic solidarity (and not anomie) was indeed the main effect of division of labor. Nor do they address the question of how that *could be* so. These questions were also not addressed by papers that appeared in the Spring 1995issue of *Sociological Perspectives* celebrating the 100th anniversary of Durkheim's *The Rules of Sociological Method.*

In retrospect, however, Ross can be seen to have regarded anomie—the disintegration of a society's normative system—as the *essential* (most often expectable) effect of the advanced division of labor so characteristic of an industrial society.[178]

Have we in fact become a society in which people's routine activities constitute, in effect, preying upon each other ?[179] The society about which Durkheim was writing did not yet have television, and an advertising *industry* had hardly begun to develop as the ubiquitous instrument of commercial predation it has now become. As Thomas Frank has suggested, today's TV advertisers and business management theorists consistently assault traditional values of loyalty, respect, and decorum. They do this, apparently, because corporate America regards expressions of cultural decay as seductive in messages aimed at selling products.[180] And continued selling of products is imperative—because with today's division of labor no specialist is self-sufficient.

If that is the nature of effective TV advertising, then "organic solidarity" has at least been seriously undermined by the inescapable side effect of modern interdependence. The ethos of modern commerce, according to Frank, "is the root cause of the unease" felt by Americans toward the culture surrounding them. Disunity and cynicism about major institutions are evidently increasing.[181] The unease has become global. The commercial ethos is international, a facet of modernity wherever it has developed. Such an ethos is the natural accompaniment of great occupational diversity—intrinsic to industrial-level division of labor. It may be especially rampant in the country where, as Calvin Coolidge famously asserted "The business of America is business," but it is by no means *unique* to the United States, or even confined to one hemisphere.

As noted above, Durkheim's book is cited in later sociological literature far more often than Ross's, but a greater prevalence of Durkheim citations than of Ross citations in the recent literature of sociology does not suffice to prove Durkheim's pro-specialization view more valid than Ross's "at one another's mercy" view. A majority of sociologists may have been committed simply by a tradition of their academic discipline to a mistaken impression of the way things work.[182]

Vested Interests and Treadmill Living

Modern societies are "becoming ensembles of social organizations that persist only because persons are coming to experience their interests, energy, and hopes as irrevocably tied to their functionary status in the service of some physical or social technique," says Manfred Stanley.[183] People are stuck on a "treadmill of production" that operates not only for most industrial societies but also even for Third World societies, according to Allan Schnaiberg.[184] While he sees this treadmill as especially pronounced in capitalist nations, in my view he has described a fundamental attribute of *industrialism*; it is by no means exclusively capitalistic. If the treadmill shows up in some Third World countries, too, I believe that it does so mainly as an unsought side-effect of their efforts to become industrial. The problem goes much deeper than the qualities of capitalism versus communism, etc. It is the same problem whether a regime's sacred scripture was authored by Adam Smith or by Karl Marx. Division of labor impairs self-sufficiency, necessitates exchange among specialties, and thus tends to cause people to treat one another as "resources" and/or "customers." We are dehumanized.

There is a continuing market for edible products because the food we eat today is digested and metabolized, so tomorrow we need to eat again. Demand for performance of the many occupational roles involved in bringing such products into existence and getting them to our tables is perpetual. But occupational categories that specialize in providing "durable goods" are different. Workers in a television factory, for instance, may have no further livelihood once every household has been equipped with a satisfactory TV set—unless having multiple sets per household can be made the norm, or new flat screen models can make possession of an old cathode ray tube TV seem demeaning.

Being "on a treadmill" means, beyond implications of drudgery or tedium, that each member of an industrial society has an abiding interest in never letting his particular job be finished. The system's need for his special function must never be allowed to become satiated. The custom of making annual "model changes" in automobiles and major appliances—so familiar to denizens of modern societies that few think to question its naturalness or inevitability—seems to exemplify this treadmill compulsion to ensure perpetual "need" for the latest somewhat altered version. Especially when it can be touted as "all new!" The need to keep on selling is almost undeniably a major factor generating "style changes" in any fashion industry—clothing, housing, whatever. Sellers of any product or service have an incentive to encourage public belief that *new* equals *improved*, the latest version is better than "last year's model." Artificial obsolescence, a wasteful custom inducing customers to replace a still functioning possession with a newer model, is a strong argument against Durkheim's optimistic appraisal of division of labor.

Moving the Merchandise

The autobiography of the man who headed two of America's "big three" automobile manufacturing companies (one after the other), contains many indications of the overwhelming tendency for one's outlook on life and on the world to depend upon one's role in the occupational web of this industrial society.[185] When Lee Iacocca was young and was forging up through the ranks at Ford Motor Company, that company's East Coast regional manager advised him, "Make money . . . . Screw everything else. This is a profit-making system, boy. The rest is frills." In the early 1950s, having been promoted to Assistant Sales Manager for Ford's Philadelphia district, Iacocca justified certain innovative selling practices with the view: "Whether or not the dealers are moving them, the cars keep coming off the assembly lines and you've got to do something about it . . . . You learn to produce, or you get into trouble—fast!" It is a further symptom of the grip division of labor has on human thought that Iacocca used the verb "produce" not to refer to making cars but to making sales. And selling (extracting money from car-buying customers) was referred to euphemistically as "moving" the cars.

Even the political affinities of this business leader were shaped by his place in the division of labor. "As long as I was at Ford and all was right with the world," said Iacocca, "I was a Republican." But when he took over at Chrysler, with several hundred thousand people about to lose their jobs from that company's impending demise, he found the Democrats to be "the ones who were pragmatic enough to do what was necessary. If the Chrysler crisis had come up during a Republican administration, the company would have gone down the tubes before you could say Herbert Hoover."[186]

In principle it is conceivable that a given society needs only so many cars, so many TV sets, or nuclear missiles, only some finite number of copies of a particular sociology textbook—or only so many sociologists. However, because academic sociologists prosper when numerous students enroll in sociology courses, it is not to our interest that society's finite need for sociology graduates ever be fully met. Either completion of our task of educating the appropriate quota, or any other change of conditions that would reduce the number of newly minted sociology graduates per year required by society, would be a threat to our own academic employment. Similarly, a textbook publisher is threatened by the consequences of market saturation, whatever text he may have published.

So it is common practice to "drive the used books off the market" by bringing out revisions more frequently than might otherwise be justified, especially if no way has been found to make the book so fascinating that its student users will just keep their copies permanently even when they have finished the course in which the book was assigned.

The world may little note nor long remember the fate of obsolescent sociologists or last year's textbook authors, but consider this: engineers, whether American or Russian, who earned their living building intercontinental missiles were tied by the division of labor system to an interest in continued stockpiling of their product even when enough was already enough. Retired U.S. Navy Admiral Eugene Carroll, who had opposed the MX missile on strategic grounds, stated bluntly, "Once you get the program rolling, the money itself builds constituency." And the mayor of Sacramento, California, in whose county the largest employer at the time was Aero jet General, builder of the second stage of that MX, expressed "fear that the outbreak of true peace would bring massive unemployment."[187]

Prospective Self-Destruction

It bears repeating. Our language ability is at the core of our *human* nature. We are the ape that speaks[188] Our original linguistic skills enabled us to invent, among many things, language in other than spoken forms, and we have gone on from learning to write to creating ways of mass dissemination of our written words. In addition to words that become literature and poetry, the language ability has given our species a vocabulary of numbers and other mathematical concepts. We have generalized our language ability and found other types of symbols to manipulate, conveying many kinds of meaning, with various influences upon human activities. There is a word that refers to a special range of significant symbols: money. Systems of monetary symbols powerfully influence modern living. It is scarcely possible to be human today and not be influenced by the central role in modern cultures of money in its various forms.

Within the past century we learned to go beyond print as a means of mass dissemination of our words: to radio, television, the internet. Language in its various forms has enabled us to diversify in cultural ways. We have piled extensive social (non-DNA-based) diversity on top of the gene-based diversity between and within human populations. So we live now with division of labor *in extremis*. And we have barely begun to understand all the implications of that fact.

Our species-wide ability to speak, calculate, write, innovate, criticize, and spread ideas has transformed the relations of each human to other humans and to the non-human world. It not only enabled humans to build elaborate cultures and to develop complex political organizations (governments of various types).

It enabled us to do science, to achieve technological progress, and transform ourselves, in effect, from *Homo sapiens* into the technology-abetted species with a colossal appetite for resources and a colossal impact upon the planet that supports us—best designated *Homo colossus*. We went far beyond the ancient command to "be fruitful and multiply" and multiplied our avarice, and our dominance in the biosphere.

Think of the contrast between the per capita appetites and impacts of chimps versus humans, a contrast much greater than the mere difference in body size, or the very modest difference between their DNA and ours. Another thought that bears repeating: the dwindling populations of chimpanzees and bonobos, our closest animal cousins, are in danger of losing their remaining crucial habitat not from *their* misuse or overuse of it, but from ours. Now that we have become *Homo colossus*, our own critical habitat (planet Earth) is in process of being rendered less fit for supporting continuation of our colossal ways of life—and that transformation is of our own doing!

Such self-destruction is not unprecedented. In earlier publications I have compared *Homo colossus* to single-celled yeast organisms thriving in a wine vat, where they initially find an abundance of the resources—moisture and fruit sugar—required for the yeast way of life. Carrying on their fermentation lifestyle (what led the winemaker to put them into that finite environment), they multiply exponentially, and not only deplete the resource base for yeast life, but their effluents (CO_2 and alcohol) accumulate around them. By transforming the finite habitat upon which they are utterly dependent, the yeast make it less and less capable of continuing to support them. The exploding yeast population thus brings about its own demise—a population crash. That procedure was not what was meant by the ancient command, "Go thou and do likewise." But our species *has* unwittingly emulated the yeast—in a big way, big because our way with symbols enabled us to become *Homo colossus* and allowed us at the same time to be improvidently prolific.

Our Predicament in a Nutshell

Arguably, at the industrial level of development, division of labor has become a markedly unprogressive force. We look upon life under preindustrial cultures not only as having been, in the words of Thomas Hobbes, nasty, brutish and short, but we see those cultures as tending to make people unreceptive to innovation. However, the high degree of occupational specialization so characteristic of our own society constitutes a comparably powerful influence that locks us onto accustomed paths, sometimes prompting us to be nasty and to act in brutish ways toward our fellow humans, and keeps us unresponsive to indications of disastrous consequences just over the horizon. As Barrington

Moore has pointed out, even some people with potential grievances are tied by a strongly sanctioned division of labor into the prevailing order.[189] What does that prevailing order do to our human qualities? "There is no work so dirty or dangerous but that it will attract volunteers pleading wife and babies to support," said E. A. Ross a century ago.[190]

The accustomed ways into which we are locked by division of labor press us toward disregarding the golden rule. Instead of doing unto others as we would have them do unto us, because of modern circumstances we are pressured toward doing whatever will get from others whatever we have been induced to want.

I do not mean to suggest that we are *all, all the time*, brutish in our treatment or even our attitude toward *all* others. I am only saying a serious undercurrent in that direction is inexorably spawned by modern division of labor. To paraphrase the familiar triptych aphorism by Abraham Lincoln about fooling the people:

> Division of labor does pressure all of the people to be somewhat brutish toward some other people sometimes.
>
> Division of labor does pressure some of the people to be brutish toward most others most of the time.
>
> But division of labor's pressure does not make all of the people brutish toward everyone always.

That third statement, however, should not be construed as *exoneration* for division of labor. The two prior statements express the fact that division of labor has an insidiously corrupting *influence* on people everywhere in the modern world.

Imbued with Mercenary Spirit

Modern methods of mass communication have spread information, knowledge, and occasionally wisdom, but their use has been perverted (by what division of labor has done to relations among human beings) and they have become a force for enslaving the public with the mercenary spirit. We all know the mass media are used for advertising and propaganda, and there have been many studies attempting to assess the "effectiveness" of propaganda campaigns and advertisements. However important it might be to know whether a particular ad has, or doesn't have, the intended effect (of selling something, presumably), it would be far more important to consider the overall impact of the deluge of mass communication on the kind of worldview people are induced to hold. That problem has not been studied in depth. There are some concepts, however, by which the *probable* effects of the deluge can be considered. Anomie is one of those concepts.

Anomie is a French word Durkheim used. It is sometimes translated as normlessness. It would be more accurately considered a condition of society in which means are not regulated by social norms as effectively as ends are prescribed by a culture's implicit values. We learn goals, then learn to pursue them by whatever means come to seem effective, with diminishing regard for standards of equity, morality and decency. Is deceptive packaging of merchandise necessary to keep sales volume up to a desired level? Notice at your local supermarket how often the transparent plastic wrappers or bags containing certain food products (e.g. oranges, or hot dogs) are printed with lines or checkerboard patterns in a color that so enhances the outward appearance of the merchandise that you can't tell even how good it actually looks until you have made the purchase, taken the bag home and taken the contents out of it. How often are the strawberries in the bottom layer in a wrapped basket as luscious looking as the top layer? Aren't the slices of bacon always arranged in the package so as to accentuate the lean and conceal much of the fat?

Anomie is a condition tending to infect human lives in any society that has extensive division of labor and has the exchange processes necessitated thereby. This tendency is aggravated in societies with mass communication media. It is easy to see the possibility that mass media as used today are instilling anomic attitudes. People are being led to want many things and are being induced to suppose that satisfying those wants may require illicit actions.

In television drama, how often is virtue portrayed as sufficient to bring desired outcomes in life situations, without any resort to violence, toughness, or deceit? How strictly is the reporting of news kept distinct from the provision of entertainment? Or to consider a finer detail of the business, what do the people in the broadcasting industry think the word "next" means? In a good dictionary it is likely to be defined as equivalent to the phrase "following immediately"—*not* as "following after something else intervening." But the infallibility of an avid TV-news viewer's taking it as a cue for immediately pressing the mute button to silence a barrage of commercials whenever the news anchor says what news story is "coming next" is on the same order of magnitude (between 98 and 99 percent) as the identity of human and chimpanzee DNA. "Coming next" in practice *almost always* means on TV not really next but "after these advertising messages."

Recognizing that division of labor eliminates self-sufficiency, makes exchange necessary, and fosters mutually predatory human relations, enables us to understand why the mass communications industries behave as if the old notion that "honesty is the best policy" has become an obsolete precept.

Pandemic Ulteriorism

It is important to see that the foregoing statements are more than just indictments of workers in the broadcast industries. Noting these aspects of broadcasting should be a warning about the ways in which the reading, listening, or viewing public has become susceptible to imbibing a destructively cynical outlook on life. "You can't believe everything you hear" too easily becomes "You can't believe *anything anyone* says," and people with that outlook simply cannot function in the long term as human beings. No society comprising a citizenry of utter cynics can long endure. Are today's "modern" societies committed to procedures that manufacture cynics?

If *Homo sapiens* is the communicating animal, we have an endemic requirement for communication—we need to be able to *rely* on information communicated to us from other people. Even non-human species that live in groups rely to some extent on information obtained not by the individual's own sense organs alone, but communicated to the individual from other members of the group. Reliance on the sense organs of conspecifics is a pattern that antedates the origin of *Homo sapiens*, but we humans had been able (until recently) to magnify that advantageous pattern. Language made possible reliance on a whole magnificent network of other people's nervous systems and sense organs.

Today's level of division of labor, insofar as it imbues suspicion that much communication is being done for ulterior purposes, weakens or destroys that ability to rely on a network of information sources. It reduces each of us to reliance more nearly on a single set of sense organs, our own. Destroy trust in verbal inputs and you destroy a core attribute of human nature. Allow your own trust in the inputs from others to be destroyed and you allow yourself to become less human. When others' words too often flow from ulterior motives, they spread the fatal disease of "ulteriorism." It is a social disease, or rather an *anti*social syndrome. insidiously destructive of human trust, sympathy, and cooperation.

Serious Instances

For a glaring example of such ulteriorism, which ought to shock even the most jaded distraction specialist, consider the way a major American television newscast began as long ago as June 5, 1968:

> *Co-anchor, Chet Huntley:* Senator Robert F. Kennedy was shot in the head and gravely wounded early today before hundreds of people in his political headquarters in a Los Angeles hotel, a month and a day after the assassination of Dr. Martin Luther King in Memphis, seconds after he had made a speech celebrating his victory over Senator Eugene McCarthy in the California Democratic presidential primary . . . .

Continuing the reporting of this event, the scene shifted to the hospital:

> *Jack Perkins*: The latest medical bulletin . . . says Senator Robert
> Kennedy remains in extremely critical condition . . . .

Frank Mankiewicz, the Senator's press secretary was then shown reading the medical bulletin. Perkins had some more to say, and then the camera returned to Chet Huntley for further reporting of certain aspects of the situation. He was followed by the face and clipped voice of Co-anchor David Brinkley.

> *David Brinkley*: . . . we have assembled some of the film from last
> night, beginning with the Senator's victory speech at the Ambassador
> Hotel, after he won the California primary.

The film, lasting several minutes, showed the speech, the cheers from the crowd, the moment of the shooting and the ensuing pandemonium and near panic, the frantic and repeated requests—"Is there a doctor in the house?"—the wounded Senator on the floor, police cars taking the suspect away to jail with crowd reactions as he is brought out and sirens fading into the distance, and then the grief-stricken crowd in the hall again. Then this:

> *Announcer*: The Huntley-Brinkley report is produced by NBC News
> and brought to you in color by Newport, the smoothest tasting
> menthol cigarette—Newport king size, and the new extra long
> Newport Deluxe 100's.

Then a filmed commercial showing a frivolous barbershop scene:

> Said a patron whose name was McNair,
> As the barber was trimming his hair;
> "This new cigarette has the roughest taste yet!
> Who's got a smooth one to spare?"
> Then up spoke a fellow named Dave
> Who had just finished having a shave:
> "Newport, you'll find, is a much smoother kind,
> With a taste about which you will rave."
> *Chorus singing*:
>   Smoother Newport, Fresher Newport—
>   Smoother, more refreshing cigarette!

This was followed by a filmed commercial for tires sold by Phillips 66 dealers, concluding with the slogan, "At Phillips 66, it's performance that counts." Then:

> *David Brinkley*: The police are holding a young man charged with the shooting . . . .

What to make of this? Apparently no senator or presidential candidate need die in vain; his assassination can, after all, attract an audience to whom such commodities as cigarettes[191] and tires can be sold. And children in the TV audience can discover that this is what life and death are all about.

But the intrusion of commercial sales pitches upon one's attention while one is trying to follow ostensibly non-commercial communication content is not confined to the broadcast media. It occurs in print as well.

Consider the following striking example. In October of that same year, 1968, the "November" issue of McCall's magazine appeared on newsstands carrying the article by the late Senator Robert Kennedy entitled "Thirteen Days." The magazine proclaimed this to be "the story about how the world almost ended." One would suppose such an eschatological story would not easily be mixed in with product-peddling hucksterism. Despite that reasonable supposition, however, it was virtually impossible to read continuously—without having one's attention diverted to juxtaposed advertisements—straight through the account of the unquestionably momentous decision-making processes by which the United States had coped with the Cuban missile crisis and averted nuclear war between the world's then two superpowers.

The sequence of actions which had secured the removal from Cuba of those weapons of mass destruction—a sequence which no single government could fully control or even predict—could not be followed by the magazine reader without interruption. The precariousness of our lives, so evident in this behind-the-scenes account of atomic-age history, was offset by the advertiser's and magazine makeup editor's skill in diverting the reader's eye from the text by conspicuous placement of full-color advertising matter.

On the title page there was a photograph of author Robert Kennedy's concerned countenance. On the first page of text there was one picture apiece of Soviet Chairman Khrushchev and U.S. President John F. Kennedy. Across the top of the next two pages were small pictures of the faces of ten other principals in the story—Cuba's Fidel Castro, General Taylor, Defense Secretary McNamara, McGeorge Bundy, Secretary of State Rusk, Soviet leaders Andrei Gromyko and Anatoly Dobrynin, Theodore Sorenson, George Ball, and UN Ambassador Adlai Stevenson. Beyond these there were no further illustrations *germane to the story*, not even any of the U-2 spy-plane photographs by which the presence of the Soviet missiles in Cuba had been established, nor any pictures of the U.S. Navy ships carrying out the quarantine.

After the fourth page of text, the reader was instructed to "turn to page 148," in the center of which, surrounded by a narrow band of text, was a full color ad showing a block of cheese and half an onion and almost seeming to trivialize the U.S. naval quarantine established in waters surrounding Cuba to intercept Soviet ships, by proclaiming "Only Saran Wrap keeps them miles apart."

On the next page there was even less text and an even larger ad, in color, urging the reader to "Invite your friends to munch . . . Chex party mix" with Planters peanuts. Turning the page, the reader's eye was again deliberately distracted by a centrally placed color ad for Neo-Synephrine nasal drops, and on the right-hand page again, an even larger ad, in hard-to-ignore color of course, for feminine hygiene deodorant spray.

Turning that page, one found three more columns of text next to one column of assorted black-and-white ads—for an anti-perspirant, an itch-relieving skin cream, a chapped lip remedy, hand lotion, a denture adhesive, and a proclamation of "cat week international" by the American Feline Society. On the facing page 153 to the right was a full-page color ad for Christmas gift subscriptions to McCall's magazine.

Turning the page again, the reader's eye might survey several more ads and read a line or two of text, finding it confusingly irrelevant before discovering he had missed the fine print instruction at the bottom of the preceding page (p. 153) indicating that the continuation of the missile crisis article was to be found on page 164, not page 154. The magazine makeup editor had not tried very hard to prevent the reader from making this error. But as the reader leafed ahead it would turn out there were *twenty* pages to get past to reach the continuation. The interval between page 154 and page 164 included ten additional *unnumbered* pages of virtually nothing but advertising, mostly in color.

Having arrived at the page numbered 164, then, the reader would find almost half of it occupied by a color ad for Kellogg's Rice Krispies cereal, including a recipe for using that product in marshmallow treats, and more than half of the next page (facing, on the right) was devoted to a color ad for fresh almonds from California. Turning to the next page, one's eye was again drawn to a centered ad in color, surrounded by a thin margin of the late Senator's ominous text, used now to help sell facial tissues. To the right, on the facing page, a color ad taking up more than half the page showed an attractive young woman reclining in a bath tub to convince the "reader" of the merits of a certain brand of bath oil.

Next page, a margin of text surrounding a graphic ad for Anacin, to relieve Nervous-Tension Headache—a malady less likely to have resulted from the missile crisis than from the would-be reader's eye-movements having been deliberately rendered chaotic, or even from such of life's problems as depicted in the ad on the facing page: "How to get to the onion salt without knocking over the Worcestershire. Get a Rubbermaid turntable for 50¢ and 2 Del Monte labels" (color picture plus coupon). Next page, a centered ad, mostly black and white, for a denture cleanser. Right-hand facing page, a large color ad for continuous action cold capsules. Last page, half text, half color display (in two parts) advertising women's shoes.

This crisis in the Cold War between the Soviet Union and the United States was one that could have been *final!* Why, the reader might well speculate, *didn't* the world end—if this magazine treatment was the way human beings must react to such events?

Cultured *Homo sapiens* has come to the point where our exchange networks apparently oblige us to use stories of episodes like that superpower confrontation that could have been final so unblushingly as bait to obtain an audience for attempts at selling products, many of them outright superfluous. Clearly there is need for research into the impact a lifetime of exposure to such mixtures of the momentous and the trivial will have on the values internalized by readers, viewers, and listeners.

But there is an immediate basis for inferring what has gone wrong with the value system of those humans who happen to operate the communication media. They employ specialists whose assigned task is to distract—make it as difficult as they can for the reader to get through an important article without having attention diverted to advertising. The function (or more accurately, the *dys*function) of these specialists is to make it nearly impossible for readers or TV viewers to learn of the day's news without having thoughts diverted to purchasable products.

No Real Improvement Since

Yes, a few limits (such as elimination of tobacco commercials from American TV) have been imposed in the years since those episodes, but the mercenary tilt remains a pervasive aspect of modern culture. In the spring of 2006 I received a letter from *Vanity Fair* magazine's Vice President and Publisher, together with a free copy of that magazine for the month of May. The letter began by telling me "Vanity Fair is pleased to announce the publication of its first Green Issue" and described it as "a special edition addressing one of the most critical topics affecting our nation and the world: the environment." The front cover of the magazine was predominantly green in color, and portrayed four assorted celebrities, all in one way or another identified with environmental concerns. They were Academy Award-winning movie actor George Clooney; Julia Roberts, film actress and also an Academy Award winner; Robert F. Kennedy, Jr., well-educated lawyer, author and environmental activist son of the late

Senator; and Al Gore, former Vice President and known for the documentary film about global climate change, "An Inconvenient Truth."

The magazine, it was said, was devoted to addressing global warming with the goal of ensuring that the topic would remain "an imperative part of the national conversation." But once again, this issue of this magazine had the excruciatingly cluttered layout emphasizing advertisements that make undistracted reading of serious text an extremely difficult task. The first item in the whole magazine that was connected with its cover-promised "theme" was a two-page article beginning on page 58 (but interrupted by a full-page ad for Krug Champagne) telling about the four-member "Green Team" pictured on the cover. It was hard to tell from the words on the two pages whether readers were supposed to be any more interested in the views of these four individuals about the ecological impacts of six billion human earthlings than about the purveyors of the four celebrity environmentalists' pictured clothing and hair styles.

Starting on page 200 with a two page photo spread, at last one found several pertinent articles almost uninterrupted by ads. The title of the first one, by Mark Hertsgaard, radio and print journalist, "While Washington Slept" was printed across the upper part of an aerial photo of Washington, D.C., modified to show that capital city seriously flooded the way it would be as a result of the rise in sea level to be anticipated from a complete melting of the Greenland ice sheet—a vast frozen body which a small caption in the lower corner of the photo says did already shrink in the previous year by 50 cubic miles. On pages 202-203 another modified aerial photo shows New York City's Manhattan streets inundated, by even higher sea levels expected (perhaps centuries hence) if greenhouse gases lead to elimination of the Antarctic ice sheet. Hertsgaard's article continues, uncluttered, just through page 206, at the bottom of which, in *very small* print are the words "Continued on page 238." A reader who fails to notice that instruction to skip the next 31 pages and just turns to the next page will there be confronted with the beginning of another article having nothing to do with the magazine's "green theme." Instead it is about "the lingering mystery" of the case of the Boston Strangler. *That* story goes on continuously as far as page 213, at the bottom of which, again in easily overlooked fine print, are the words "Continued on page 234." Not noticing that instruction to skip twenty pages, the reader who just turns one page is confronted with a two-page nude photo of Golden Globe-winning film actress Keri Lynn Russell. What could this have to do with either global warming or the lingering mystery?

Will Anyone Bell This Cat?

Rewind the tape of modern human history, back to Saturday, the 30th of June, 1860. We look in on a session of the meetings of the British Association for the Advancement of Science that has been convened in the long west room of Oxford University's Gothic revival museum. We hear purple-vested Bishop Samuel Wilberforce orating a lengthy squelch of the implications being drawn from Charles Darwin's newly published theory of natural selection. Among those listening is the man who would come to be called "Darwin's Bulldog," Thomas H. Huxley. In the course of the Bishop's two hour string of eloquent phrases expressing righteous disapproval of the idea of evolution, he turns to Huxley and asks him whether the apes from which he claims to be descended were on his grandfather's or his grandmother's side.[192]

Huxley responds by declaring, "If . . . the question is . . . would I rather have a miserable ape for a grandfather or a man highly endowed by nature and possessed of great means of influence and yet who employs these faculties and that influence for the mere purpose of introducing ridicule into a grave scientific discussion, I unhesitatingly affirm my preference for the ape."

Fast forward to the present. Where is there now such an intelligent, rational and quick-thinking member of our species who, like Darwin's Bulldog, is ready to rebuke societal "standards" that have grown so degenerate? Must we permit insidious pressure toward mercenary interactions among people to produce results on the level of the foregoing examples of mass media usage? Should the precious symbol-manipulating ability of our species be so wastefully (and destructively) deployed? Was that our destiny? Has the division of labor and the invention of money done irreversible damage to the promising hominid line?

Couldn't we somehow have done better?

CHAPTER 9

How Life and Foresight Are Possible

"In Newton's day it was generally understood that the ability to predict future positions of the moon could be of significant help in solving the pressing problem of finding longitude at sea."

—I. Bernard Cohen, *The Newtonian Revolution, p. 30*

"It soon transpired that the eclipse had scared the British world almost to death; that while it had lasted the whole country, from one end to the other, was in a pitiable state of panic, and the churches, hermitages and monkeries overflowed with praying and weeping poor creatures who thought the end of the world was come."

—Mark Twain, *A Connecticut Yankee in King Arthur's Court*

In the centuries that have elapsed since Newton devised his impressive explanation for the regularities of heavenly motions, and even in the time since Darwin showed how the concept of natural selection explained patterns he observed in the diversity so evident in Earth's plant and animal kingdoms, we the symbol-wielding species have learned many things and have extended our vocabularies. But we are still sometimes misled by words we use too uncritically. In this chapter, therefore, we will explore some of those words that tend to mislead. This may seem at first a digression from views of the human situation developed in previous chapters, but the reader is urged nonetheless to "stick with it." Be assured, the conceptual clarifications to be developed below will be indispensable for understanding ideas in subsequent chapters, and especially for understanding the future toward which our civilization is hurtling.

Prediction Happens

In baseball, an outfielder often is seen starting *at the sound of the bat* to run toward where he will catch the high fly ball.[193] His subsequent success in making the putout is almost routine, but it is possible only because he has developed a skill in quickly observing the very earliest segment of the ball's trajectory and has also sufficient power of mentally extrapolating that initially ascending curve. He intuitively converts the extrapolation into knowledge of where on the field, and when, the descending baseball will arrive at a catchable height from the ground. When the ball smacks into his glove (and is not dropped), the batter is out—and the outfielder's "mathematics" have succeeded.[194]

It is important to remember that whenever a fly ball is caught, the forces that impelled it to the location where the catch was made were imparted by (a) the motions of the pitcher, and immediately thereafter by (b) the impetus of the batter's swing, and then (c) movements of the air through which the ball was passing, and (d) gravitational attraction between two masses, ball and Earth. The fielder's movement toward the place where the ball could be caught did not cause the ball to come to that spot. The significance of this simple idea that causes come prior to effects (even in the realm of evolution) will be evident in the final chapters.

Obviously, most humans cannot catch batted fly balls even remotely as well as major league outfielders, and probably few spectators in the ballpark will think that the players are performing *mathematical* procedures (other than mere counting of strikes, balls, outs, hits, errors, and runs). But most of us humans do have—to a much lesser extent—*a similar kind* of ability. Almost any two people can "play catch," tossing a ball back and forth to each other. One person need only toss the ball in a direction that brings it *near enough* to the other person. Throwing ability need not be so precise as to project the ball exactly to where the catching person's hand happens to be at the instant of throwing.

Catching a thrown object involves some ability to foresee where it is going to be after the time lapse required to bring it from person A to the vicinity of person B (will it arrive above or below B's line of sight, how much to right or left?) so B's hands can be thrust out as may be appropriate for making the interception.

As a pastime, this simple game of toss-and-catch is possible only because two very fundamental things are true:

---

(1) A thrown ball's trajectory does follow a path through the air that *accords with physical laws*—the curve is calculable, thus future locations are predictable.

(2) Our nervous systems have a trainable capacity for eye-hand coordination.

Even for those of us who are not professional ballplayers, being able to predict future events to a reasonable extent, and being ready to respond at the appropriate time and place, is important in any life.

Prediction serves many species

We humans are not unique in our possession of nervous systems with predictive capabilities. One's pet dog can also catch a ball tossed near it, not with hands, of course, but by coordinating in much the same way certain other quick muscular efforts, using an automatic mental extrapolation of even a *bouncing* ball's motion, so that the dog's open mouth arrives at a location on the ball's trajectory at the same instant the moving ball reaches that place. (Shnop! Got it—commence tail wagging.)

A fish may not seem as smart as a dog, but there are some fish whose survival depends upon their ability to foresee the trajectory of a moving object. When an archerfish (*Toxotes jaculatrix*) sees an edible insect on foliage hanging above the surface of the water, it accurately squirts a water jet strong enough to dislodge that prey. The insect tumbles forward and down on an approximately parabolic path. Archerfish have no vocabulary containing the adjective "parabolic," but several such wordless fish are likely to be swimming around together in a shooting party, so the shooter is unlikely to consume the waterlogged insect it shot down unless it can arrive ahead of the competition at that bug's landing place. Research has shown that an archerfish appears somehow to "calculate" where a plummeting fly is going to land. A few ballistic parameters determine the point of impact. The fish appears to "rely on visual measurements," which it makes in about one-tenth of a second—of the initial height and horizontal velocity of the insect. Thus it is able to locate its prey's eventual landing spot and swim immediately to it. What is unclear, however, is whether archerfish actually calculate each trajectory from scratch or learn some average approximation from past experience.[195]

There are also birds with such a skill—so precise, in fact, as to make a baseball outfielder marvel. They are even *called* flycatchers. They routinely catch their insect prey in flight.[196] This means they have the ability to foresee where a flying insect's path through the air is taking it and the ability to alter their own flight path quickly and sufficiently to achieve timely intersection with the path of the moving prey. Their skill is somewhat more impressive than the ball-catching dog's or the stunned-bug-catching fish's, for the bird must foresee the probable motion of a living and evasive creature (a bug unimpaired by any water jet). The bird's living target has a capability of making midcourse corrections in its own flight path, a capability not possessed by either the archerfish's squirted target or a thrown or batted baseball.

In short, with respect to moving objects (animate or inanimate), we and many other creatures have the ability to *predict*. We, in fact, might not even be here to consider these instances of reaction to predictions had our prehistoric ancestors not had enough bird-like ability to foresee the probable motion of the prey animals they hunted. They were able to relate to that movement of meat on-the-hoof and estimate reliably the direction to throw a spear or shoot an arrow. Our lives depend on predicting—which, in turn, depends on *the fact that many events and processes are predictable.*

Just the kind of universe we need

Prediction is possible because *we live in a predictable universe, with discoverable regularities among its many phenomena.* Only in such a universe were the achievements of minds like Newton's and Darwin's possible. At the human level, with our evolved capacity for using symbols, we have enhanced this predicting ability through mathematics, so that we can even predict future locations of celestial bodies! Astronomers extrapolate the calculated movements of objects in the sky. Thus it is possible to "foretell" the arrival of, say, a comet into the region close enough to Earth to be seen with the naked eye. People read an announced prediction in their newspaper and venture outdoors on a prescribed evening to look in a specified direction to behold a fuzzy spot of light not seen there on previous evenings. Or we journey to an announced location on a specified date to watch an eclipse of the sun, because the orbital movements of Moon around Earth and Earth around sun have been precisely calculated. Similar calculations have enabled us to send spacecraft to intersect the orbital movement of Moon, Mars, or Venus, to engage in picture taking fly-bys of the larger outer planets and their moons, or to purposely collide with a designated comet or asteroid.

Being in a predictable universe is what makes life possible.[197] We embody much of its predictability in our own lives. Predictions enable us often to do things that work as we intend them to, that attain ends sought. Sometimes this attainment of goals (with modern devices, especially) may seem like magic. But the predictions, the actions, and the predicted outcomes are not magical, and the fact that they are not magic is important, although in ordinary experience that fact is often clouded.

As mentioned earlier, one of my memorable experiences as a little boy was attending a world's fair in Chicago in the 1930s. The fair was called "A Century of Progress," as it had been organized in defiance of the economic hardships of those Great Depression years, to celebrate Chicago's centennial. My childish perceptions of various technological wonders displayed at the fair doubtless differed immensely from impressions left with most adult visitors. One item I remember was a "magic" drinking fountain. Water began flowing from it only when one bent over it to take a drink—when it marvelously "turned itself on" without being touched by its user!

As those who contrived it intended, this seemed like magic, but my parents explained to me how it was done. There was a photoelectric cell on the rim of the fountain bowl and light beamed from the ceiling above to that "electric eye" was interrupted by the head of the person bending over the fountain to quench his or her thirst. Interruption of the light beam stopped an electric current that was keeping a hidden valve turned off, so the valve went to its open position, allowing the water to flow.

Few kids my age, if any, could have constructed such a system, even if they'd been handed all the components, but it was nevertheless a mechanism a child could understand when given that explanation.

I remember my own slightly perverse reaction. Wouldn't it be fun, I thought, to reverse the circuit so that water flowed *until* someone bent over to take a drink, automatically *shutting off* in response to the person's interruption of the light beam. Both the way that fountain was actually rigged, and my imaginary reversed arrangement, were simple cause-effect systems. Modern life is filled with mechanisms by which human wants are more or less reliably served. Occasionally, however, if something fouls up these mechanisms they frustrate rather than serve.

Intended outcomes *can* fail to occur. When that happens we are frustrated unless or until we learn why. For some malfunctions, we sometimes derive almost as much satisfaction from learning the reason for their occurrence as from having them put right. But failures of the mechanisms in our lives can be costly. It can also be costly to misconstrue either success or failure, to embrace wrong explanations of how they happen. We often hold obsolete knowledge, and persist in reacting to successes or failures in ways that vitiate the desirable outcome or aggravate the failure.

Of course I only imagined and never actually constructed a reversed circuit automatic drinking fountain, but perhaps it was poetic justice anyway that many years later, as an adult I owned a typewriter that developed a bothersome flaw and was repeatedly frustrated when using that device. One of the links inside it often came loose. It was a link between the machine's Q-key and the lever whose type bar was supposed to strike the ribbon and print the letter Q on paper. That one detached link didn't seem to affect any other links. But when the Q link came loose, pressing that key accomplished nothing. Fortunately most of the words I was likely to write did not include that letter, so the typewriter was not altogether useless. Sometimes, when my thoughts did involve a Q-containing word, knowing of the defect I could devise a circumlocution to evade the problem (e.g., substituting "silence" for "quiet" or "interrogate" for "question"). But not always, and anyway having to think of suitable synonyms wasted time.

It was occasionally quite necessary to use some Q-containing word, without evasion, and I would then have to interrupt my writing and spend from 10 to 30 minutes unscrewing some screws, removing the machine's cover, and poking into its innards with an old-fashioned button hook to try to reconnect the detached linkage.

I was fortunate, of course, that it happened to be Q rather than, say, E that had this fallible linkage. Since E is the most-used letter in ordinary English prose, if the defect had been located on the apparatus for printing *that* letter, my typewriter would have been thoroughly useless. But even with only sporadic interruptions pertaining to the less frequent letter, the typing experience was sometimes quite infuriating.

So naturally there were occasions when that internal linkage in my typewriter came apart and my distress erupted in vocal expression. It is still all too human to get angry at a recalcitrant helper, even when the helper is just tool or a machine. But of course, whatever the content of my vocal outburst, it was promptly followed by my silent recognition that my spoken exclamation, no matter how vigorous (and "justified") it was, had no influence on the inanimate typewriter. Only by opening it up and poking with that old-fashioned tool could the unhooked link be reconnected.

Reliability versus magic

The point is that our use of typewriters represents a common human experience of *relying* on an artificial mechanism that dependably links actions to outcomes. We expect an electric lamp to come on when we turn the switch. We expect the car to slow down or come to a stop when we depress its brake pedal. We expect a familiar voice to answer when we dial the phone number of friend or relative. We expect the TV channel to change when we press appropriate buttons on the remote control device.

Any such working mechanism is an example of a *deterministic* system. By that I do not mean everything is *predetermined*. Some people use the word "determinism" as if that were what it must mean. I do not. Certainly no suggestion is intended that "come hell or high water" I would necessarily dial specific numbers and talk to, say Fred Smith, at 10:21 a.m. on Tuesday. What I do mean is simply that in our use of common mechanisms such as typewriters or dial telephones the action-to-outcome connection is generally consistent, dependable, and understandable.[198] A world that is deterministic in that sense enables science to be done, leading to scientific explanations.

Explicit understanding of *how* pressing a particular key puts a particular letter onto the paper eludes most of us most of the time. We are satisfied merely to know that the thing does work and generally care not how this is possible.

Cursing a mechanical failure may serve an emotional purpose but it doesn't put things right with the failed device. Whether or not one considers oneself a "determinist" in a broad philosophical sense, we do live in a culture which tells us denouncing a machine is ineffective as a means of repairing it.

The opposite of determinism is magic. What if the universe operated in such a way that when I emitted words of exasperation I had then immediately found I could type Q after Q after Q? Our real world is not one where that could actually happen. To imagine it happening is to imagine a *magical* repair. Magic supposes outcomes need not depend upon clear and adequate (and ascertainable) causal actions. Otherwise put, belief in magic imputes causal sufficiency to acts or processes that are in fact not sufficient to cause the particular outcome.

Mark Twain provided a clever fictitious account of "magic" based on what he called "transposition of epochs" which enabled an American who went to sleep in 1879 to awaken in unfamiliar surroundings which he was told were in 6th-century Britain.[199] The superior 19th-century knowledge of the deterministic solar system available to this displaced Connecticut Yankee in King Arthur's era enabled him first to test the veracity of his informants' claims about the date, and then to stun them and their countrymen with an ostensibly magical demonstration of his power over celestial events. As he puzzled over his predicament and the unbelievable information about where and when he had awakened, he happened to recall that the only total eclipse of the sun in the first half of the sixth century occurred on the 21st of June, A.D. 528, at 12:03 p.m., which was just a day or so hence (according to the local calendar). He also knew there was no total eclipse of the sun due in what to *him* was the present year—1879. So, just prior to the appointed time he threatened to blot out the sun unless his captors changed their hostile intentions toward him as a stranger. Of course as the scene darkened when the eclipse happened, to them it seemed both a fantastic performance by him and a serious threat to everyone's future. The reader, however, knows it was no threat and had only the *appearance* of magic.

On a more mundane scale, we also know that a verbal expression of outrage at a machine's malfunction is not an adequate causal action capable of rectifying its performance. Most of us tend to presume we have outgrown in the real world reliance on magical incantations as means for attaining ends. The fact that we wouldn't really expect mere utterances to restore the faulty typewriter's proper functioning reflects prevalent contemporary acceptance of the idea that magic is only imaginary.

Even so, we enjoy reading about imaginary worlds in which people get things done by magical means. We enjoy seeing magical actions portrayed by the "special effects" crews in the movie industry.[200] And we occasionally enjoy watching "magicians" perform either on stage or doing parlor tricks. But we know enough to refer more accurately to such entertainers as *illusionists* and to wonder *by what deception* they achieve the *appearance* of magic.

Putting aside illusion-based "magic" (perhaps we should call it pseudo magic), what would constitute "real" magic? If someone were thought to possess magical power, it would have to be an ability to achieve an intended result by acts that any reasonable human being ought to consider inadequate or ineffectual—to fly by a sheer act of will, say, rather than by using some kind of aircraft. A world of magic would confront us with many unexplained *and inexplicable* events—happenings achieved by sorcery rather than by ordinary skills.

Now suppose one sits down in front of an intact and properly functioning mechanical typewriter with a sheet of paper properly inserted and positioned for writing. Five keys are pressed in succession: m-a-g-i-c, and the word "magic" appears on paper as intended. Being familiar with the typewriter, we do *not* consider the word's appearance on the paper a magical occurrence. Even those of us who lack detailed knowledge of the typewriter's inner structure (and therefore less than fully understand how it enables one's fingers to achieve the intended result) usually regard it as a deterministic mechanism which straightforwardly converts our alphabetic thought into the typed word. If we think about it (and have at least some vague notion that there are pivots and levers inside) we may consider the accomplishment impressive—but not miraculous, not magical. If the system did not involve deterministic connections between a writer's actions and the resulting appearance of symbols on paper (i.e., if it did not perform reliably) it could hardly be useful.

Determinism, Fate, and Choice

Now by these terms it follows that *a human organism is (normally) also a deterministic system.* An adult human body happens to be a system so constituted (and so trained, by the culture of a modern society) that intentions conceived in the brain can reliably send through certain nerves the messages that reliably cause muscles in one's arms and hands to do the motions that depress those typewriter keys that spell out intended words. Earlier human brains would have directed arm and hand muscles to manipulate pen or pencil to achieve equivalent (but not identical) results—less swiftly, and with less precision.

Suppose one sits down in front of an IBM *electric* typewriter, again with paper appropriately inserted, and presses in succession four keys: f-a-t-e. There is a soft humming sound from its electric motor as the little metal "golf ball" rapidly swivels, rotates and nods, striking the carbon ribbon. By means of raised type that adorns its surface (letters, numerals, punctuation marks and a few mathematical symbols) this little sphere does the impressing of designated letters consecutively upon the paper, spelling out the word "fate." This is a more advanced typewriter, but it responds to the actions of the writer precisely as it was constructed to do. Just as with the earlier machines, the predictability of its conversion of key pressings to typed letters on paper is what it means to call its actions deterministic.

Two versions of typewriter, very different inner mechanisms, both instruments for putting words on paper, equally deterministic in their performance. Neither typewriter committed magic, in mechanically converting finger pressings of keys to a word on paper. Nor was either machine *fated* (or predestined) to type either word; it only did so because its human operator *chose* to type those five or four letters.

Fate and determinism are not synonymous, and both differ from magic. Neither has to contravene the fact that humans choose to act one way or another. Which is not to say human choices are uncaused. The causes underlying human choosing are seldom simple, nor are they uniform even within a family, let alone among the various members of an extensive population. The causes behind our choices, being manifold and diverse, tend to remain sufficiently invisible that we commonly construe our choosing, even when statistically somewhat predictable, as uncaused. Our free choices are not random. They are not uncaused. Each word one writes is partly determined (i.e. influenced) by the words preceding it—as well as by words that will follow, all pertaining to thoughts forming inside one's head. A person uses a typewriter to "write" words on paper in a language he or she has learned from the surrounding culture. Modern typewriters (and computers) can be an improvement over writing with pen or pencil, both because they can be faster—more nearly synchronous with a writer's thought processes—and because the result is more reliably legible. But a typewriter that becomes unreliable is hardly preferable to a pen or pencil.

Pens and pencils were available in societies with less elaborate division of labor than ours involves. People in those societies had fewer advantages and more disadvantages than we have. As I have acknowledged in previous chapters, the very extensive division of labor so characteristic of our modern industrial society has enabled helpful devices to be mass produced for us to obtain and use. As I've said, "If every author had to grow his own wheat and bake his own bread to obtain the strength to mine his own ore and smelt his own iron to make his own typewriter before starting to write, the world would have few typewriters and few books (even on agriculture, baking, mining, or manufacturing)."[201]

Specialization, exchange, and breakdown

Division of labor is elaborate and essential in modern societies, but as previously noted it can be advantageous only when associated with a functioning system of exchange of products and services among the numerous specialists. Adam Smith was so impressed with the magic-seeming operation of the system in which each specialist pursuing his own specialty supposedly somehow harmoniously served the collective good that he likened it to a process being directed by an "unseen hand." But to be so advantageous this exchange system has to be *non*-magical. It must have near-mechanical reliability. It may occasionally tolerate minor malfunctions (equivalent to that unlinked Q) but *utterly* self-serving actions by society's members could not *ensure* system survival, much less system harmony. If unreliabilities become too pervasive, malfunctions can devastate human societal life—as we saw in 2008 when national and international "financial systems" began collapsing. It took vast infusions of money by various governments to begin putting this monetary Humpty Dumpty together again. Even former Federal Reserve chairman Alan Greenspan had to recant his long-held faith in dependence on the magically virtuous functioning of bankers' self-interest.[202]

Many interconnected human skills have brought these "writing" devices into existence, but the fact remains they are inanimate. They only put words on paper when caused to do so by the actions of fingers, directed by a conscious brain, which actuate predictable operations of the mechanism inside one form of typewriter or another, each form being a deterministic system in which finger actions, through one kind of linkage or another, result in specified letters appearing on paper.[203]

Now, still other devices exist for converting thoughts into writings. Consider, for example, my Macintosh computer on which *this* writing was done. It has a keyboard, somewhat more elaborate but basically similar to that of a typewriter. This keyboard sits between me and a flat screen. I press in turn the keys m-a-g-i-c, and watch those letters appear on the screen, spelling out the word I intended. No levers connect these keys to the screen, but each key pressed sends an electric impulse through a cable into the computer's marvelously mysterious electronic innards, and by means of this enormously *different but no less deterministic* system the word my brain told my fingers to contrive duly appears on the screen. Equally deterministic auxiliary mechanisms enable me to have words on the computer monitor's screen reappear as inked text on paper inserted in a connected Epson printer.

I no longer need to use a button hook for making repairs to keep on typing, but insofar as my ignorance of the innards of the computer is even more abysmal than my very primitive comprehension of the insides of those older typewriters, the appearance of meaningful language on the screen and the process of transferring documents from screen to paper via the ink-jet printer does *seem* like magic. Ours is a culture, however, in which computers have been abundant for a number of years, and knowing that electronic technicians can correct some malfunctions that do occasionally occur—or even give me step-by step instructions by telephone for doing minor corrections myself—I know these "magical" effects are *not* magic. The computer, too, whatever its wondrous complications, is a deterministic apparatus.

Regarding it as an "electronic brain" made learning to work with it offer insights into things we do with the brains in our heads. For instance, experience with it can clarify and emphasize the meaning of determinism as I'm using that word. Suppose the computer were somehow fouled up, so that pressing the key for a particular letter merely put up on the screen not *that* letter but just any random letter, so that no matter how often I tried to type f-a-t-e, the computer insisted on putting an utterly random selection of four letters onto the screen, say "xbmo," then "tnfg." *That* would indicate the normally deterministic linkage between actions of fingers on keyboard and appearance of designated letters on screen was inoperative. Such a fouled up (non-deterministic) computer would *fatally* obstruct my determination to write anything intelligible in the way of a discourse contrasting the concepts of determinism and fate.

No less truly, if the brain-to-fingers messaging inside a human would-be writer were disrupted by an injury, illness, or stroke, the ordinarily deterministic system of that writer's bodily apparatus would malfunction. For normal human beings, malfunctions of that sort are sufficiently uncommon that writing is a feasible task, and civilization has advanced.

Thankfully the task of writing is greatly facilitated by the mechanical and electronic devices called typewriters and computers invented by collaboration among many variously-specialized humans. By further collaborations among many other variously specialized humans, such devices have been abundantly manufactured, sold, and bought. These intricate gadgets are, in effect, exosomatic "organs." Thinking of these devices as extensions of our "bodily apparatus" makes it easier to recognize the deterministic nature of our own bodies, comprising nerves, muscles, bones, etc. If my brain were disrupted by, say, Alzheimer's Disease, or my arms were paralyzed from injury or infection, my ability to press keys on a keyboard and write meaningful paragraphs could be lost. The truth of that statement attests to the merit of considering the human organism as a system that is deterministic when whole and healthy. Like other mechanisms its deterministic operation is susceptible to disruption. It may be a truism to say that a deceased human can no longer type, but it is no trivial observation. It reminds us of the similarity of human bodily apparatus and the prosthetic extensions with which we equip ourselves in modern times. Computers can "crash;" we can die.

Antonyms, Not Synonyms

Language and concepts function for us insofar as they, too, involve reliable connections. But those connections can also become disrupted. Suppose I am a writer in good health, whose fingers obey cerebral commands quite accurately, but suppose also I have internalized a misconception of determinism or of fate, so that I construe the two words as synonyms. Many people do. That confusion of terms would tend to impede my correct comprehension of the way things work in the real world.

To show how, let us suppose that no matter which four keys I press on my computer keyboard, and no matter the sequence in which I press them, the four letters spelling *fate* always appeared on the screen. From one point of view that would be a most exasperating malfunction. But it could also be seen as a small but disturbing exemplification of the concept of fate. Fate is a word we use to refer to outcomes that happen regardless of anything we do, regardless of contrary hopes and intentions and efforts.

Human attitudes about concepts like fate vary. We tend to admire people who resist the temptation to "resign themselves to their fate." Yet we also sometimes see wisdom in "going with the flow."

A tendency to misconstrue determinism as fate may preclude awareness of quasi-magical expectations that linger despite exposure to a culture that generally embraces science and disavows magic. Let us be clear about these terms:

Determinism means there are reliable *and discoverable* connections between events, between actions and results, causes and consequences.

Fate usually means an outcome cannot be influenced by any purposive action a person or a group might take. Intentions cannot alter *pre*determined outcomes.

Magic would mean one could obtain a desired outcome without undertaking the actions ordinarily necessary and sufficient to produce it. In a world of magic, mere wishing, or wishful incantation, might suffice to produce the result. It was only by importing into the Britain of King Arthur's time various items of American knowledge and technology achieved over thirteen subsequent centuries that Mark Twain's Connecticut Yankee was able to maintain a better-than-Merlin reputation as "magician."

Clarifying the Human Situation

We do not live in a magical universe. Nor is the course of events in our world so predestined—so fated—that it is uninfluenced by what we do.

Is it reasonable to dream of living in a world of magic? Perhaps, but life would be harder, not easier, if the dream came true. Magic is useless—actually worse than useless, because it would block perception of reality. It could also be worse than merely useless if the magic were being done by a thoroughly evil person. We should rejoice that magic is unreal.

Magic connotes sorcery. It refers to events being brought about by actions that could not have sufficed as causes in any ordinarily deterministic world. There cannot be magic in a world of deterministic devices and processes. If this is a world of deterministic processes and devices, there cannot be magic. Magic implies supernatural (or non-natural) causation—as if automobiles could be produced not by the actions of machine tools and a labor force of diverse engineering and mechanical specialists in an automobile factory but, say, by a qualified wizard chanting "let there be automobiles."

Ironically, it is most likely our experience with using deterministic devices to serve *chosen* purposes that misleads us into denying determinism. We recognize that others might have chosen to make different uses of the same devices. But what we really mean by denying determinism is that what the devices do for us is not *pre*-determined (prior to our own desires and goal oriented actions). We expect the devices to do for us what our actions "tell" them to do. So we require devices that *are* deterministic, *in that sense*. Fate, on the other hand, is a concept denoting that our actions are *not* reliably connected to outcomes, that some events and circumstances will happen regardless of anything we may do.

Fate usually connotes unpreventable *and often unwelcome* circumstances. Perhaps because we sometimes can use foresight to prevent unwanted conditions from coming about we reject fatal-*ism*. That rejection misleads us to imagine we have to reject *determin*-ism, as if fate and determinism were the same. They are not. Determinism and fatalism are virtual *antonyms*, not synonyms at all. Determinism is a concept denoting that causes and effects are reliably linked and that the linkages are in principle discoverable. Fatalism is a belief that events are *pre*-determined regardless of anything we may do. In rare instances that belief may hold, but it is not the general picture of life in the real world.

We are rejecting fatalism (but *embracing* determinism) when we check the car's fuel gauge before driving out of town—to forestall the unpleasant experience of running out of gas "in the middle of nowhere." And we do refuel when needed. We also get the car "serviced" on schedule, so that it will reliably perform (as a deterministic mechanism).

Blurring Future Prospects with Word-Magic

Sometimes undesirable events do happen despite (inadequate) human efforts to prevent or avoid them, so the notion that we are subject to fate, much more than we wish, is understandable. Cultural advancement and technological progress are supposed to have enabled us to become less subject to fate than in earlier times. Insofar as this is so, it is because we have learned more of the predictable connections in nature between cause and effect, so our avoidance efforts more commonly succeed than formerly. For instance, knowing how various life-threatening diseases are caused, we sometimes learn how to take curative or even preventive actions, and prolong longevity. We have made real progress in numerous sectors of life. Having learned to make handy transportation facilities, we can move resources about, and reside and work in places on our world's surface that would otherwise have remained uninhabitable. Learning the many things electrons can do when they flow through circuits of many forms in ever more diverse contrivances, we can "be in touch" with people over great distances.

Desirable outcomes also sometimes happen even when we do not understand the processes that led to them. When good things happen, and we don't know what particular overt action was taken to bring them about, it can seem miraculous. There may even be a *hint of magic* when the favorable results of actions deliberately taken to bring them about happen also to be accompanied by certain rituals. The golden words or ritual behaviors may have no real connection with the "result"—but they may acquire "*gilt* by association." A ritualistic word like "presto," mouthed just prior to pulling the rabbit from the ostensibly empty hat, augments the illusion that the rabbit was conjured from thin air. But apart from that defining trick of the professional illusionist, we are all susceptible to a kind of conditioning process that enables us to confuse, as causes of results, the words or other symbols that happen to accompany whatever effective actions really produced the results.

Conversely, sometimes it's not the uttering of "magic words" but the sorcery of silence that we practice. A mildly pro-magic attitude among baseball aficionados manifests itself for example in a social mandate to *withhold* comment when a pitcher has got through several innings without giving up any base hits. When the opposing batters through the first several innings of a game have been "set down in order," it is taboo for the radio or TV play-by-play announcer to mention explicitly the possibility of the pitcher achieving (by continuing this for the entire nine innings) a "no-hit game"—as if speaking the words would jinx him.

The same sort of taboo seems to occur in more serious realms. For fear of "spooking the market" (and giving bargaining power to OPEC), oil consuming nations in the early years of the 21st century hesitated even to *discuss the possibility* that the peak of world oil "production" might be imminent.[204] Yet the soon-to-come transition from ever-increasing to inexorably declining availability of petroleum and products made from petroleum will have to bring on a period of truly momentous social change in societies that have allowed themselves to become so fossil-fuel dependent. Refraining from discussing this will neither prevent nor even postpone it. When that transition happens, it will not have been caused by verbalized foresight. *Lack* of verbalized foresight can aggravate the societal impact of that change, but leaving the issues unspoken cannot increase the amount of petroleum that was formed and accumulated underground long ago by pre-human geological processes. Any of us who may have been naïvely receptive to such word-magic delusions must forthwith abandon them.

To some extent, conventional notions of entrepreneurs can be confused with notions of magic. The art of *financing* the production of something is sometimes construed as magic-like—insofar as we think about it in overly abstract terms, using a vocabulary that focuses on the availability and use of money and does not explicitly take into account the material resources required, the energy needed, etc., apart from the role of money in arranging their use.

Solid citizens who would today deny any belief in magic nevertheless allow themselves to be misled by faulty vocabulary, as when the extraction of nonrenewable resources from the Earth is referred to by the same word used to denote farm or factory output—*production.* Other conventional words mislead as well. Energy industry people, for example, know that not all of the oil discovered is going to be technologically or economically "recoverable." But why is the extraction of *some* of the finite stock of a fossil fuel called "recovery"? Neither the coal nor the petroleum that exist in underground locations where nature put them are something we once had, then lost, and are now getting *back*. Recovery's first syllable (the two letters *r* and *e*) endows that word with a connotation of return to a former presence, when in fact, the fossil fuel was laid down in the earth's crust long before humans even existed, let alone knew of its combustible qualities or its very existence. When we speak of fossil fuel extraction not only as "production" but also as "recovery," do these vocabulary habits impel us to assume the stuff *belongs* above ground where it is *meant* to be burned (or, in some instances, converted by molecular rearrangement into plastics or solvents or other humanly useful forms)?

Is the stuff "useless" until we get it above ground? Or was its ancient burial by the processes of nature a blessing in advance which we "sapient" humans, to our peril, have failed to recognize? Now, when concern has arisen over the long-range effects of industrial civilization's combustion products—atmospheric change causing an increased "greenhouse effect" leading to serious global warming and more severe weather—an enormous irony occurs among suggested "solutions." Carbon dioxide is identified as the chief culprit, the most abundant of various "greenhouse gases." So there are proposals to "sequester" the carbon dioxide that results when carbon-based fuels are burned. When those biological and geological processes long ago removed so much carbon from Earth's primordial atmosphere and locked much of it away underground, they were *already doing* an enormous sequestering job. Without nature's removal of those vast quantities of carbon from the atmosphere, Earth would not have had the kind of air, climate and surface environment in which animal life could evolve. There could not have eventually come along *Homo sapiens*, with impressive (but not unlimited) powers of comprehension.

Perhaps the ultimate human error was our recent ancestors' innocent misconception in defining (and thus perceiving) the underground carboniferous substances as available *fuels*. Perceiving those substances that way was a consequence of the technological developments that were beginning to convert *Homo sapiens* into what I have called *Homo colossus*[205]—people with gigantic power, with magnified per capita resource appetites, and enormous environmental impacts. Unquestioned definition of the sequestered carbon as fuel-meant-for-burning led to massive, on-going, ultimately disastrous *desequestration*. In just over two centuries, fossil-fueled industry wrought such transformation of human societal life that it seemed magical. But it was also transforming the planet, and especially its atmosphere, upon which human life depends. Now, as the mixed blessing of industrial magic has become more ominous, some people are winsomely proposing to evade dire consequences by putting some of the genie's offspring—the abundant CO_2 molecules—back in the underground bottle. To assume it is somehow feasible—by re-sequestering the oxidized carbon—to extract *only the energy* from the naturally sequestered hydrocarbons and expensively avoid befouling Earth's atmosphere with climate altering by-products of the energy-extraction process, may turn out to be misconception no less monumental than that previous misdefining (it burns, so *of course* it's fuel for us to use)—the definition that has already so changed our world.

How Fates Happen

Undesirable outcomes do happen even when we do not understand the processes that led to them. When bad things happen, and we fail to recognize any overt action taken that might have brought them about, it seems like fate.

But fate can happen from more than one kind of causal process. Fate is an outcome that literally couldn't *not happen* under the prevailing circumstances. But circumstances that we humans collectively have brought about will have consequences we never intended, and not having foreseen the outcome it will *seem* like fate. As people have made themselves more colossal, it has become more important than ever to recognize that there is more than one way in which a future situation can be unpreventable. We need to distinguish between what may be called *absolute fate* and other instances that should perhaps be called *quasi-fate*—or *anthropogenic "fate."* (The latter phrase is only an oxymoron if we omit the quotation marks around "fate.")

Absolute fate. Suppose Earth were to collide with another asteroid like the one which is believed to have ended the reign of dinosaurs in this planet's animal kingdom, and to have extinguished a majority of the other life forms then extant. Effects of the impact would presumably wipe out human civilization, and would probably bring extinction to many species, possibly including *Homo sapiens*. That would be fate, in the sense that nothing we had done, and nothing humanly doable that we left undone, would have had any connection to the extinction of ourselves and other species. Nor would extinction by such a catastrophe signify a lack of *fitness*—for successful living in the pre-collision environment. The old notion that dinosaurs became extinct because they just overdid their success and grew too big was misleading. In their time, for the ecological niches then available, the size of each dinosaur species was adaptive. But the whole Earth changed too abruptly for slow processes of evolution to fit most of their potential descendants to post-collision emerging niches, so extinction was the dinosaurs' fate (except for the small, bipedal dinosaur species enough of whom survived to become ancestors of birds).

But what will *seem* like fate for humanity in our "advanced" era (barring any catastrophic intervention by an extraterrestrial body) can, in principle, come about in one of two ways, two kinds of quasi-fate. First, we, or our customs and institutions, can be overwhelmed by the *cumulative unintended results of purposive actions*. Or, second, we can collectively attempt to head off a catastrophe but find ourselves doing unhelpful or even counterproductive things or, simply, *too little too late*.

Anthropogenic "fate." More than ever, as we move onward into life on a planet whose human carrying capacity has been exceeded by an ever growing population of ever more technologically colossal humans, we must invoke one of the keenest insights of a passionately concerned and unusually popular sociologist, the late C. Wright Mills. It was an insight by which he sought to assist his contemporaries in perceptively reading the news of their times. We will need to become more perceptive in order to avoid misconstruing events that are going to be happening in the years ahead. It will be helpful to have understood "fate" as Mills saw it. He discerned and neatly defined the first type of quasi-fate—the ultimate cumulative impact of unintended side-effects of purposive actions.

One of his most earnest books transcended archaic thoughtways enough to note that only sometimes and in some places do humans *make* history; at other times and places, the minutiae of everyday life can add up to what he said was mere "fate." Even though Mills wrote from a paradigm that was pre-ecological, he was able to provide an unusually clear conception of this important condition.[206] "Fate" (at least of the type I have decided to call anthropogenic "fate"), is shaping history, Mills explained, when *what happens to us was intended by no one and was the summary outcome of innumerable small decisions about other matters by innumerable people.*

It is by such manmade "fate," that we humans have come to be living unsustainably[207] in a world with insufficient carrying capacity to support permanently six-plus billion of us—especially if we all become as technologically colossal as Americans (which the people in China and India, numbering in the billions, seem definitely intent upon doing).

As we encounter the effects of a worsening *carrying capacity deficit* we shall be sorely tempted to assume some person or group must have intended to bring about whatever bad things are happening. But when adverse events or circumstances have resulted from *anthropogenic "fate"*—"fate" in the Mills sense—then blaming, resenting, and retaliating against someone can be both futile and dangerous. We must learn to see the deep-seated temptation, so characteristic of most members of our species, to blame and retaliate, as an expression of a lingering but woefully obsolete tendency to resort to *magic*. Scapegoating, after all, is a form of magic. And assigning culpability to some putative "enemy" in response to "fate" is blind scapegoating.[208]

It was in precisely Mills's sense of "fate" that a marvelous carrying capacity surplus was converted into a competition-aggravating and crash-inflicting deficit after two centuries of the spread of industrial activity and human proliferation on this finite planet. Thinking more ecologically about human history than Mills was equipped to do, we can see that the predicament confronting humanity was the summary result of all our separate and innocent decisions to trade horses or oxen for a tractor, to avoid illness by getting vaccinated, to move from a farm to a city, to live in a heated home, to reside far beyond walking distance from our place of work, to buy a family automobile and not depend on public transit, to specialize, exchange, and thereby prosper, even to reproduce beyond replacement level.

We will need to keep in mind that to a very large extent the horrible aspects of life in a future beset with a deepening carrying capacity deficit will have come about because of things *almost everyone* hopefully and innocently did in the past. It wasn't just decisions and actions by someone in particular that inflicted grievous circumstances upon us, and subjected us to horrendous experiences. To single out supposed perpetrators of our predicament, resort to anger, and attempt to retaliate, will be the ultimate folly. Our acts of indignation will cause unforeseen outcomes all too often seriously counterproductive. They will compound the ostensible fate. The misfortunes these acts will inflict upon us will thus be a second sort of anthropogenic "fate"—*aggravated "fate."*

There were a few brilliant minds who foresaw the possibilities. Global climate change is a surprise to most people, but the idea that human burning of coal and petroleum, by injecting increasing quantities of CO_2 into Earth's atmosphere could have a "greenhouse effect" was recognized nearly two centuries ago.[209] At first it remained simply unknown to nearly everyone not connected with chemistry, physics, or atmospheric science. Later, as an increasing number learned of the idea, many shrugged it off as a matter of little real concern. This seemed a big world, too big to be much changed overall by little human beings. As more knowledge pertinent to the matter accumulated, though, early skeptics began to see that we *could* actually change the world. Then, whether this recognition was cause for alarm, or grounds for mere fascination, or even for hopeful anticipation, depended largely on what seemed likely to be the effects of climate change on one's particular occupational specialty (or "vested interest"). National leaders who might, with adequate foresight, have embraced policies (like the hopeful but less than adequate Kyoto plan) to avert or postpone disaster, insisted we were not really headed for disaster.[210]

If catastrophe happens from the cumulative atmospheric injections of CO_2 from all the use we and our forebears and immediate posterity have made of "fossil fuels" in pursuit of comfort and the enjoyment of assorted products of industry, the "fate" that overtakes us will be from too little preventive action having been taken too late.

We can make it worse. We will do so if we stubbornly misconstrue events and attempt what we think is "remedial" action by *retaliating*, as we shall be overwhelmingly tempted to do, against people we perceive as villains who supposedly caused our troubles.

CHAPTER 10

How Carrying Capacity Limits Freedom

"Whereas once humankind exerted its will in the relatively small arena of artificial selection (the arena I think of, metaphorically, as a garden) and nature held sway everywhere else, today the force of our presence is felt everywhere. It has become much harder, in the past century, to tell where the garden leaves off and pure nature begins. We are shaping the evolutionary weather in ways Darwin could never have foreseen; indeed, even the weather itself is in some sense an artifact now, its temperatures and storms the reflection of our actions. For a great many species today, 'fitness' means the ability to get along in a world in which humankind has become the most powerful evolutionary force."

—Michael Pollan, *The Botany of Desire: A Plant's Eye View of the World*, 2001

Freedoms have ecological limits. Although this fact has been addressed before in various ways[211] it is *not* uniquely a human issue; animal studies have illuminated it. For example, India's population of tigers, one of many species much reduced in the 20th century, has been relegated to fragments of its former habitat so limited in extent that "population pressure" among these animals has changed their behavior. Relations between tigers, and between them and the human inhabitants of adjacent lands, have changed, and both species have lost some former freedoms in this process.[212] Such studies deserve both public and academic attention.

The aim of this chapter is not to worry about an endangered feline predator but to indicate some ways in which human freedoms depend on our own changing relation to the human carrying capacity of the world (i.e., its maximum sustainable human load).

As wisely noted years ago by the Russian-born sociologist Pitirim Sorokin,[213] freedom "becomes restricted when [a person] cannot do what he would like to do; has to do what he would prefer not to do; and is obliged to tolerate what he would like not to endure." The world's peoples are still practicing cultures that give them desires that may be unfulfillable in a condition of carrying capacity deficit, a term I shall explain below. Expectations persist that are becoming increasingly unrealistic in a world where the relation between human load and carrying capacity is changing adversely. Thus, there is ecological meaning in British anthropologist Malinowski's statement that "freedom of action means the conditions sufficient and necessary for the mastery of all circumstances inherent in the execution of purpose . . . ."[214]

Among the freedoms subject to ecological restriction are at least two of the "Four Freedoms" proclaimed in 1941 by Franklin Roosevelt and Winston Churchill in The Atlantic Charter as worth fighting the Second World War to defend. It ought to be self-evident that "Freedom from want" is subject to carrying capacity limits, but the link between "Freedom from fear" and carrying capacity may be more subtle. Other freedoms, too, as we shall see, are curtailed or threatened by a carrying capacity deficit.

From Population Pressure to Load Pressure

More than three decades ago, a book by a British sociologist, Jack Parsons, on a topic close to the subject of this chapter, suggested that all intelligent and reasonably well educated people in our time should be aware that their own future is endangered by overpopulation.[215] "Overpopulation" should of course be recognized as a relative term. In the human case it means too many people—but in relation to something else. Since it connotes a ratio, it can be understood by considering a shrinking denominator no less than by focusing on a burgeoning numerator. As the tiger example showed, some effects of range compression may be indistinguishable from effects of population increase.[216] Man-caused scarcity of tiger habitat intensified competition among the tigers (and occasionally converted some of them into man-eaters) no less effectively than would an actual influx of additional tigers onto a fixed amount of land.[217] Among humans, to speak of overpopulation endangering our liberty is to speak of freedom being imperiled by "scarcity."

To pursue the point we must ask: scarcity of what? I shall try to show that our species, too, is beset with an equivalent to the experience of range compression. This is happening to us at the same time that our numbers are increasing. In this chapter I will argue that it is a deficit of carrying capacity relative to the human load that is our fundamental problem. I shall need to define and explain very clearly the carrying capacity concept, which is too often ill-understood. First, however, a major amendment I must make to the thesis of that British sociologist I quoted involves recognizing the fact that "human load" is a concept that has another dimension besides population. To understand the situation with which our species must now try to cope we must think not just of a people-to-land ratio; we must learn to think of a load-to-carrying-capacity ratio, and be mindful of the load concept's additional dimension. Counting or estimating the number of tigers gives an adequate measure of the load to be borne by a finite remnant of tiger habitat. Census enumeration of people does not suffice to measure the world's human load, for people differ from one another in ways tigers do not.

Prodigious increase in the number of people inhabiting this planet has been a key feature of life for my generation. There are over three times as many humans alive today as there were when curiosity first led me, as a boy, to ask some adult how many people there were in the whole world. Even my own country's population has more than doubled since I first read a census figure, although we were supposedly already into the final (non-growth) phase of what demographers have called "the demographic transition" before I had ever heard of that concept.

But it is less widely recognized that expansion of load on its other (non-population) dimension has at least as much to do with the problem of scarcity and consequent erosion of freedom in our present and future. The world is being required to accommodate not just more people, but effectively "larger" people—ones whose resource appetites and power to degrade the planet that supports us have been enlarged by technology.

Living organisms necessarily make withdrawals from and additions into their environments in the process of living.[218] For each person living on this planet at the beginning of the industrial era, there are today more than six people, all making these ecological withdrawals and additions. With today's technological equipment, the average daily consumption of energy by *each* of those six individuals is more than five times what was consumed daily by each of their ancestors less than two centuries ago.[219] Energy-use per capita is one good indicator of the amounts of substance withdrawn, transformed, and added back into our environment by our life activities.[220] So, what most truly represents the enlargement of the global human load is the product of the two increases—our six-fold increase in human numbers *multiplied by* at least a (global average) five-fold increase in per capita energy use. Thus, since 1800 the load imposed upon earth's ecosystems by humankind has grown to be *at least thirty times* as large today as it was on the eve of the industrial era (roughly eight generations ago). The question of global human carrying capacity seems far more urgent when we think in terms of "load pressure" on the world's ecosystems instead of contemplating mere "population pressure."

Industrial and Non-Industrial Loads

For some parts of the world, load has grown mostly along the population dimension. Those are the areas we call underdeveloped (or developing) countries. Elsewhere more growth has occurred on the per capita resource-use dimension, and we speak of developed countries. Growth on either dimension can intensify load pressure. It may overshoot carrying capacity. In a world of limited carrying capacity, freedoms of people everywhere are then necessarily curtailed by further growth of either sort.

From the first United States census (in 1790) to the most recent one (in 2000), enumerated population rose from 3.9 million to nearly 300 million. In 1790 for each American there were, on average, almost 57 hectares of national environment (roughly equivalent to 115 football fields) in which to live, from which to extract a living, and into which life's end-products could be deposited. By 2000, despite territorial expansion, this per capita share had shrunk to about one-fourteenth as much, or an average of a little over 3 hectares (between 6 and 7 football field equivalents) per person.[221] However, unlike the tiger which is no larger today than its 1790 ancestor, industrial man is ecologically a "larger" animal than pre-industrial man, as data to be cited below will show. So 3-plus hectares per capita now has to provide far more and absorb far more than did each 57 hectares in 1790. It should be no surprise, then, that life in modern industrial America has to be more regulated (less free) than life in 1790 America.

At a time when ecological circumstances thus heightened the need for regulation, it was ironic that in the last decades of the twentieth century there occurred a surge of de-regulation.[222] Instead of fearing environmental damage, what was dreaded was overseas flight of environment-damaging industries to escape regulation.[223] The reason the response to increasing load pressure has been so shortsighted will be made clear in the next chapter.

To see what makes industrial society citizens "larger animals," consider the load increase in terms of its other (non-demographic) dimension. The estimated daily energy use of the average 1790 American was about 11,000 kilogram calories; by 1980, with more exosomatic extensions of his being (i.e., more technological apparatus supporting his way of life), the average American per capita daily energy use had grown to about 210,000 kcal. With the technology prevalent in 1790, an average American was the energy-using equivalent of a good sized dolphin, but the 1980 American, with more advanced and abundant apparatus to shape his living, had become the energy-using equivalent of *nineteen* such dolphins, or one *full-grown sperm whale*! To see how this might constrict freedom, envision a pod of 1980 "sperm whale Americans" confined to an aquarium only one-fourteenth the size of the one in which their 1790 "dolphin ancestors" swam.[224] Quite obviously these enlarged animals in such reduced habitat must often be unable to do what they would like to do, must often have to do what they would prefer not to do, and must be required to tolerate much that they would like not to endure.

Or, consider the following facts: India has over three times as many people as the United States, and just over one-third as much land area. With Indian population density thus about nine times that of the U.S., it is understandable why some Americans automatically think "India" when they hear the phrase "population pressure." Considering instead the more cogent concept, "load pressure," the picture reverses. Not long ago there were 120 times as many motor vehicles in America as in India while America's petroleum reserves were only ten times India's. These motorized extensions of Americans' anatomy had appetites for both fuel and space. With their vehicles, Americans expected to "consume" more space than Indians expected to do. For a given amount of space, load pressure is greater for a given population of people-in-automobiles than for the same number of people-on-foot.

It is ironic that the engineering solution to this load pressure problem, the *limited access* highway (straightforwardly termed a "motorway" in Britain) is called a "freeway" in America. Motorists who travel on "freeways" are *not* free to ride bicycles on them, nor can they enter or leave the road wherever they might desire. Such traffic arteries can only be entered or left at designated on-ramps and off-ramps (which may occur only at intervals of many miles). To facilitate swift, dense traffic, and to minimize the risk of injurious or fatal collisions, these multi-lane divided highways were designed to *deprive* us of freedom to move as flexibly as we used to do.

Moreover, the ratio of vehicles to oil reserves in America was already, a generation ago, twelve times the corresponding ratio for India. Thus, in terms of the technology component of the load concept, the load-to-carrying-capacity ratio appears to differ in quite the opposite direction from the population density difference between these two countries. Indian motor vehicles were not going to exhaust Indian petroleum as soon as American cars and trucks would use up the oil under U.S. soil. Fewer Indians than Americans had cars so fewer were "free to drive," but if other things were equal, those Indian motorists could retain that petroleum-based freedom longer than Americans would.

According to demographers, even when "momentum" was already going to take the world population past the six billion mark, growth of the human load along the demographic dimension had begun to decelerate. But that did not mean the problem was going away. World *population* was not decreasing. Its *rate of increase* is what had begun decreasing.

On the technology dimension, load continued to grow prodigiously. Lumsden and Wilson compared the past three centuries' growth of scientific knowledge with "a colony of rabbits, proliferating and doubling in size every few years." With perhaps four of every five scientists who ever have lived being alive today and "churning out new ideas and information," they wrote, scientific knowledge "doubles every ten years, faster than human beings are increasing."[225] Much of it results in commensurate growth of technology.

One further point about the unprecedented expansion of the human load in the industrial era: the radius over which human activity can adversely affect the environment is greater than ever before. Modern routines of life, however local they may seem, can have worldwide ecological impacts. Destructive climatic change from "the greenhouse effect" will not afflict just the industrial parts of the world from whose fuel combustion comes the major fraction of the CO_2 additions to the world's atmosphere; it is world climate that is changing. If I remain free to use certain substances produced by modern technology for my convenience it will not be just a portion of the atmosphere's ozone layer *over my head* that will open up and let in more cancer-causing ultraviolet light; consequences from locally advantageous activities may harm a *global* environment.[226] Today, the freedom of inland farmers to use chemicals can impair the freedom of others to live by exploiting marine fisheries, because fertilizers and pesticides leached from the land are taken by "our" rivers into "the world's" ocean. The freedom of downwind people to enjoy verdant forests can be destroyed by rain made acidic with airborne compounds vented from factory and power plant chimneys in upwind countries; the winds disregard international borders. The Chernobyl nuclear power plant mishap interfered with the freedom of Swedish, Polish, Scottish, and many other non-Ukrainian people to sell or eat vegetable or animal products grown hundreds of kilometers away from the reactor site.

Any prospect for solving the ecological problems entailed by industrial ways of living must depend on the degree to which people and decision-makers *everywhere* perceive mankind's new global reach.[227] Political or corporate boundaries (within which people have heretofore been comparatively free to disregard outsiders' approval or disapproval of their daily routines) are, ecologically, increasingly obsolete. The idea that freedom from constraint can neither be bestowed nor preserved by such boundaries is far more true and salient today than in 1623 when John Donne, English clergyman poet and essayist, to express his universalistic outlook, wrote: "I am involved in Mankinde." Industrial development has infused new ecological meaning into his declaration that "No man is an Iland, intire of itselfe; every man is a peece of the Continent, a part of the maine."[228] In all three aspects (population growth, technological enlargement, and extension of impact) increased load pressure has reduced freedom to be "independent."

Carrying Capacity

The claim that freedom is imperiled by increasing load pressure draws validity from application of a clarified concept of carrying capacity to the human situation. To define carrying capacity for this purpose, it is enlightening to work from the following axiomatic principle:

> For any kind of use of any particular environment there is a rate or amount of use that can be exceeded only by reducing the subsequent suitability of that environment for that use.

To proceed from this principle to a definition of carrying capacity, we need first of all to recognize that living organisms always live by using some environment. Typically, each species has its own characteristic way of living (i.e., of using an environment). Living always involves the user making withdrawals from, and additions to, that environment. Certain kinds and amounts of withdrawals and additions are ecological characteristics of each user type. Thus, each species requires an environment whose features enable the necessary withdrawals and additions to be continually made. The "habitat" of any given species of user is an environment with suitable attributes for the life of that species. The amount of use of a given kind (i.e., by a given kind of user) that a particular environment can sustain indefinitely (i.e., without reduction of suitability as the use goes on) must depend on the rates of renewal of whatever attributes on-going use would continually tend to change.

Too little tiger habitat for the existing population of tigers could result in the suitability of the habitat to support tigers being progressively degraded by overuse. The tiger-preferred prey species in that habitat, for example, might be "harvested" faster than they could replace themselves. The supply of prey for the predators to consume would thus decline. The reason for the decline would consist in the fact that the tiger *load* was in excess of what that environment could "carry." Tigers becoming man-eaters were practicing what economists have called "resource substitution," a concept some economists have offered as panacea for any supposed resource shortage (or carrying capacity deficit).

Now the definition:

> An environment's carrying capacity is its maximum persistently supportable load.

It should be obvious that the same territory will have a different carrying capacity for tigers than for, say, monkeys. Two such different user species will use the same environment very differently, so its capacity to keep renewing the attributes required by one species will, in general, differ from its capacity to renew what is needed by the other. However, the monkey and the tiger are each so fixed in their respective ways of using their habitat that the maximum persistently supportable load of either can be expressed simply as the number of individuals of that species whose total withdrawals and additions are compatible with the environment's rates of renewal.

With these things in mind, it can readily be seen why wildlife managers and herdsmen have understood carrying capacity as the maximum *population* that a given environment could support indefinitely. One finds an extensive literature in which carrying capacity is defined that way[229]—as the number of, say, sheep or cattle that a tract of pasture land can support without being damaged by overgrazing. But now it should be clear why our more abstract definition is important: it subsumes the herdsman's concept as a special instance, but it allows for application to the human case where an important species characteristic is extreme cultural diversity. By defining carrying capacity in terms of maximum *load* (rather than population) that an environment can permanently support, we allow for the extremely varied technologies by which different human groups use an environment. One monkey makes about the same withdrawals and additions as another monkey of the same species. Likewise, one tiger is ecologically nearly equivalent to another tiger. But the withdrawals and additions of industrial humans are vastly different from the withdrawals and additions made by pre-industrial forms of human life.

Overshooting Carrying Capacity

Now let us focus on that word "persistently" in our definition. Any given environment can temporarily support more individuals of a particular species than it could support indefinitely. In other words, a user population can, and sometimes does, increase until it "overshoots" the carrying capacity of an environment. But when that happens, overuse will reduce the environment's suitability for that kind of user population. When a load becomes excessive, in other words, it begins to destroy some of the carrying capacity upon which it depends, which eventually must diminish the load itself. (This is not made less true by our inability to estimate with assurance or precision what the maximum permanently supportable load may be. Ignorance of the magnitude of an environment's carrying capacity cannot protect freedoms from being eroded by load pressure.)

Vivid examples of animal and human populations overshooting carrying capacity and suffering the consequences have been carefully documented.[230] Range compression, as we have seen, can be an alternative route to the same consequences. Recognizing now that carrying capacity can be overshot by growth of technology as well as by population growth, we can state two corollary definitions:

1. **Carrying capacity surplus**—the condition (before overshoot or habitat loss) in which carrying capacity exceeds load;
2. **Carrying capacity deficit**—the condition (after overshoot or habitat loss) in which load exceeds carrying capacity.

For *any* user species, life must be different when there is a carrying capacity deficit from what it would be, or was, when there was a carrying capacity surplus.[231]

It is probably in those terms that we may best hope to understand much contemporary social change, including erosion of freedoms. In general we should assume that more freedoms are feasible the greater the carrying capacity surplus. Conversely, the greater the carrying capacity deficit, the more freedoms are likely to be lost (or traded away for some chance at surviving that deficit).

That kind of tradeoff has been known to happen in human history and some see it happening today. In the next chapter we will consider examples where freedoms are lost in attempts to increase carrying capacity to match existing loads. Then we shall look at instances where deficit reduction was sought by load reduction.

CHAPTER 11

How We Became Entrapped

"Ecohubris ... the most dangerous idea in the history of human thought ... intoxicates us, dampening our knowledge and sense of reality. It leads us to ... a delusionary [extravagance in resource consumption]."

Lee Freese, *Environmental Connections*, 1997, p. 238

"[Earth] is a self-contained and isolated unit .... We need to realize ... that there is no ... other planet to which we can turn for help, or to which we can export our problems. Instead we need to learn ... to live within our means."

Jared Diamond, *Collapse: How Societies Choose to Fail or Succeed*, 2005, p. 521

The economies of ancient Egypt and Mesopotamia, of some other parts of the eastern world, and of the pre-Conquest civilizations of Mexico and Central and South America have been explained in terms that boil down to this: they were economies dependent on agriculture that was limited by rainfall being insufficient or too intermittent. To accommodate a growing human load the limit somehow had to be raised. Construction (by massive but primitive labor) of elaborate irrigation systems enabled imported river water to bring carrying capacity of arid regions up to the level of their populations' need. Sometimes, too, cycles of flood and drought called for artificial drainage systems. The result, according to Wittfogel, was intense and lasting enslavement, in effect, of both the huge construction crews and the peasants, by a despotic state that could organize and manage these indispensable public works.[232] Harris has labeled this deficit-coping system "the hydraulic trap."[233]

Today, when industrial societies fear energy scarcity as the limiting factor, some writers believe we may be moving into a "nuclear energy trap" with comparable potential for restricting individual freedom.[234] Nuclear advocates argue that, unlike "soft energy" solutions to the new carrying capacity deficit (small scale, locally generated power from solar and other renewable sources), we should commit ourselves to massively and centrally generated nuclear power. But critics note that this would entail acceptance of an organizational structure able to manage nuclear's risks—in transporting hazardous fuels and wastes and in storing the latter for very long periods. Such a "solution" would put the general public under domination by an industrial elite.

Industrial societies (and *industrializing* societies such as China) have also been catching themselves in a coal trap. As the "need" for ever more energy is felt, and as the more convenient hydrocarbon fuels (oil and natural gas) become less abundant and more costly, there is overwhelming temptation to revert to burning prodigious quantities of coal, regardless of climate-changing effects of the immense release of CO_2,—and thus to mining ever more of it, despite hazards to underground miners, or extensive damage to land areas by surface mining.[235]

Turning to the load reduction alternative, one glaring example of what this could involve is China's effort to curtail the freedom of families to produce more than one child apiece, a policy the Chinese government applied to the more than one-fifth of humanity residing in that one country—as an effort to cope with recognized carrying capacity deficits.[236] Although at present many people outside China see no distinction between such a restriction of reproductive freedom and unconscionable despotism, limiting human reproduction to less than replacement is an adaptation other nations must soon begin to emulate. Throughout most of the world, people have heretofore had freedom to reproduce at will. We have also had "freedom to invent," and the two freedoms together have enlarged load, as we have seen, on both the population and the technology dimensions. In short, we have used these freedoms "not wisely but too well" (to borrow the words of Shakespeare's Othello). We were thereby creating a dilemma for posterity—who will be living in a time when people who choose to continue proliferating must condemn their descendants to technologically more modest lives. As traps go, that may be the ultimate.

Also reflecting the load reduction way of coping with carrying capacity deficit may be some of the examples of racism today. One population denies certain freedoms to another population of a different race, as a means of limiting load. For example, whatever its shortcomings, the British Empire once prided itself on hoping to evolve toward a Commonwealth of democratic nations in which something like a worldwide common citizenship for all, regardless of race, would apply. It hasn't worked out that way. The United Kingdom, with population density comparable to India's (and with load relative to carrying capacity worsened by far more "technological enlargement"), has felt obliged to restrict immigration. Immigration to the UK even from the Commonwealth is less free than formerly, and distinctions are made between the mainly white Dominions and the mainly nonwhite newer Commonwealth nations. To quote again Jack Parsons, the British sociologist cited earlier:

> It would be more desirable and morally better in my judgment if the British were not prejudiced on such matters as race, if we had room for anyone who wanted to come here, and, more generally, if everyone was free to go anywhere at any time. None of these conditions obtain, however, and this seems to be less rather than more likely to be the case in the foreseeable future as population pressure increases. We can afford these refinements of morality and spirit only in a situation of spaciousness and plenty. If we define free migration, regardless of race or creed, as one of the liberal values, we have already seen how these can suffer under the pressure of numbers . . . .[237]

Freedom to "go anywhere at any time" has even been curtailed on the world's oceans as the carrying capacity deficit's effects have become manifest. Where once there was a "three mile limit" of the extension of national sovereignty offshore, now much of the oceanic surface of the globe is enclosed in national 200 mile "exclusive economic zones" wherein foreign competition is more or less excluded from use of fisheries and other marine resources.[238]

It was predicted that as the sense of urgency about the carrying capacity deficit becomes stronger among the industrial nations, aggressive military pursuit of exclusive control over vital resources in the Persian Gulf, South America, Australia and the Indian Ocean, and even Antarctica and the Southern Seas, could get out of hand.[239] Sociobiologists and ecologists find animals more likely to claim an exclusive territory by threats of fighting when there is a short supply of a vital resource.[240] Human response to severe load pressure will tend to show some clear parallels.[241] Resource wars, in fact, have already commenced.[242]

A Look Ahead

But are we, in fact, correctly interpreting the present and future human situation by seeing a carrying capacity deficit? Lacking easy measurement of carrying capacity, can we really know at a particular time whether the human load has surpassed it or not? One ecologist (Garrett Hardin) simply turned around the axiomatic principle—that only by incurring environmental degradation can carrying capacity be exceeded—and has suggested using evidence of environmental degradation as an indicator of excessive load.[243] Instances of global environmental degradation previously mentioned should now be construed as indicators that carrying capacity has indeed been exceeded. More such evidence accumulates every year.

But first, it helps to consider an implication from simply rephrasing in negative form the axiomatic principle from which we developed our definition of carrying capacity:

> For any kind of use of any particular environment there is a rate or amount of use that *cannot* be exceeded *without* reducing the subsequent suitability of that environment for that use.

In this form, the statement can be recognized as an "impossibility theorem," in the same class with such scientific propositions as "there can be no change of momentum in one mass without an equal and opposite change of momentum in some other mass" (Newton's third law, rephrased). That way of viewing the carrying capacity concept led Herman Daly to say in 1979: "The starting point in our thinking about development should be an impossibility theorem, namely, that an American high-style high-consumption standard for a world of four billion people is impossible."[244] (Now that our number is approaching *seven* billion, the impossibility should be all the more obvious.) In other words, on a planet of limited human carrying capacity, freedom to advance from the dolphin-equivalent to the whale-equivalent level of resource-consumption, or freedom to industrialize, cannot be universal for as many people as there already are.

Load already does exceed carrying capacity. This carrying capacity deficit condition of our world will powerfully reshape the human future. Looking first at the energy-consumption side of the freedom to live the industrial way of life, we find that "the peak of the oil mountain, or the time when oil supplies are clearly beginning to dwindle because new discoveries are not keeping pace with demand, had been estimated at approximately 1995."[245] If not already in 1985 (when that estimate appeared), then at least after 1995 we could have expected some type of load-reduction process to loom as a major facet of human life. Loss of previously taken-for-granted opportunities will be part of this process, for many recent lifestyle characteristics are based on abundant energy availability.

Not just fuel *scarcity* will constrict human actions though. Freedom to burn coal, even while coal remains abundant, will be threatened by awareness of the environmental damage it can do. Use of coal injects rain-acidifying SO_2

as well as prodigious quantities of heat-trapping CO_2 (and some radioactive materials) into the global atmosphere.[246] Already by the latter 1970s there was evidence of an increasing "greenhouse effect" on our planet's climate,[247] and some weather scientists were convinced human activity was causing it.[248] It was already recognized that any remedial action undertaken would certainly entail major changes in energy policies.[249] Humanity would be less free to generate energy in vast amounts by present methods. Derivatively, many people in many places will lose their freedom to do many of the things they have been accustomed to doing.

The pace at which global climate is changing has recently been found to be faster than previously supposed.[250] The clear relevance of the phenomenon as an indicator of excessive load, or carrying capacity deficit, is reflected in the fact that serious minds have begun to question whether the earth will remain habitable.[251]

Turning attention to *renewable* resource use, here too there are indications of carrying capacity having already been exceeded. Timber, for example, has been used faster than it grows—for fuel, paper, and building material. Deforestation on watersheds in many parts of the world caused many instances of human dislocation. Freedom of world shipping to pass between the two largest oceans has been put at risk by deforestation-caused heavy sedimentation and the possibility of insufficient water in the dry season to fill the Panama Canal locks, and the freedom of Panama City's residents to drink clean water has been jeopardized.[252]

In agriculture, evidence that per-hectare yield tends to vary *inversely* with farm size has been cited as reason to regard land reform that encourages small scale family farming not just as a political or economic issue but as a way to maximize land's human carrying capacity.[253] International efforts to stop "desertification" are occurring in response to the fact that, indeed, deserts are advancing.[254] Their advance is at least to a substantial degree man-caused, so it serves as another indicator that human load has overshot global carrying capacity. Human use of the land is already destroying its usability. Not only freedom from want is at stake; freedom of farmers to make individual land-use decisions, unregulated by authorities more responsive to global concerns, cannot long prevail.

Managed Resource Allocation

We noted in the previous chapter an equivalence between the effects of habitat loss and effects of population increase. Now we can put this in more general terms: similar problems arise from load pressure due to carrying capacity reduction and from load pressure due to load increase. There have been episodes in human history in which peoples had to cope with temporary loss of carrying capacity (or loss of access to carrying capacity) on which they formerly depended. This can result from natural hazards; it also commonly occurs in wartime. The equivalence we have noted now means we can look to such episodes for possible guidelines toward coping with today's and tomorrow's load excess.

Before 1943, pre-partition India depended on annually importing about a million tons of rice from Burma. Wartime Japanese occupation of Burma cut off this supply. The British administration in India "remained faithful to the laissez-faire attitudes"—to which the British government had adhered in Ireland in similar famine circumstances a century earlier—with results that included severely rising prices for rice in Calcutta and death by famine for about one and a half million people in Bengal.[255] Elsewhere in the world, when wartime situations cut off access to formerly available carrying capacity, *departures* from free trade patterns were taken so as to avoid these dire effects. Rationing (of food, and other scarce but essential commodities as well) was commonly instituted. It worked. The lesson for the future: carrying capacity deficit requires managed allocation; "free trade" can aggravate disaster.

Managed allocation has been practiced, despite ideological disinclination, when the need for it became undeniable. Within Britain itself, at the same time Bengalis were starving, a "points" rationing scheme which left considerable freedom of choice to consumers did hold them to a necessarily Spartan standard, within which it was said that the poorest ten percent of the population probably became better nourished than before the wartime rationing. Mothers, children, and invalids received priority rations of milk and eggs.[256]

Many other European countries, belligerent or not, established rationing programs. Occasionally there were black markets, sometimes there was hoarding, which aggravated certain wartime scarcities, but despite imperfections the rationing schemes generally succeeded as means for coping with lack of access to former carrying capacity (or coping with increased demands for certain commodities by allied nations or due to military mobilization). The Soviet Union allowed prices for certain nonessential commodities to rise on the free *kolkhoz* market, but rationed essentials and froze their prices. The rationing system was described as "ruthlessly biased in favour of workers who 'over fulfilled' the production norm set for their job."[257] In the United States, ration stamps were issued to all civilians for foods and for gasoline, and, based on demonstrated need, certificates were issued for items like tires. The rationing and price control programs succeeded in restricting demand, conserving scarce resources, and ensuring reasonable equity of distribution.[258]

In Nazi Germany, there was both rationing and conscription of labor (a kind of "rationing" of human effort, to ensure its priority application to the most essential tasks). When the war ended, most of the Nazi rationing schemes and even the labor conscription had to be continued by the occupation forces.[259] Despite $1.5 billion worth of American and British relief shipments, conditions in devastated Germany remained so difficult that three years after the war the official ration for a civilian adult was only 1550 calories daily.[260]

There have been other incidents of rationing since World War II. At the time of the Suez invasion in 1956, Britain and France had to institute rationing to cope with an interruption of oil supplies. Reduced fuel oil consumption shortened working hours and caused unemployment in Europe. Some countries banned Sunday and holiday motoring.[261]

Following the Arab oil embargo in 1973, and again just prior to the renewed oil supply uncertainties from the 1979 revolution in Iran, the idea began to emerge that petroleum rationing in the United States would become necessary as a way of making the transition to a "sustainable-energy era" from America's long-established habit of prodigal dependence on this nonrenewable resource.[262]

Foraging to Farming to Foraging

In one column-inch on an inside page of the Sunday paper in the community where I live there recently appeared the following boxed advertisement:

> TIRED OF HIGH
> GAS PRICES *call, email, write, fax our*
> *Congress Members & President*
> DEMAND ACTION NOW!

That's all it said. Its implicit assumption that these officials might know what action would be truly remedial, and that their "action" could correct a problem implicitly defined simply as "high gas prices" reflects the abysmal lack of public comprehension of what has been happening to the human situation. Consider, then, the fundamental aspects of the situation so seriously misdiagnosed:

All the various hominid species that evolved prior to *Homo sapiens* were foragers. They were hunter-gatherers. So were the earliest *Homo sapiens*. Nutrition that sustained them was obtained entirely from naturally occurring vegetation or naturally available animal populations. The numbers of such hominid foragers would have been limited by the amount of sustenance that happened to be available according to processes of nature not under their control.

Thousands of years ago, however, some members of our own species discovered certain kinds of animals useful to humans could be domesticated. When people had contemplated that achievement and came to perceive even "the Lord" as their "shepherd" it was plausible to presume a more sustainable future: "I shall not want."

Someone likewise discovered ways to *arrange* for the growth of certain kinds of edible vegetation. As a result, some people became herders and gardeners. They became "ecosystem managers," in effect, assuming a totally new role in relation to local portions of the biosphere. This was a new expression of the competitive relation between *Homo sapiens* and other consumer species. Instead of relying entirely on taking what they needed from naturally occurring local ecosystems, as their ancestors (and all earlier animal species) had done, these innovators undertook to control the species composition of local ecosystems in ways that would ensure provision of grains, fruits, or flesh that could serve as sustenance for humans. Humanly managed local ecosystems provided some measure of sustenance-security for localized human populations—at the expense of other consumer species which might have depended on whatever former products were now locally made *less* abundant.

For some ten thousand years following "the dawn of agriculture," the ecosystem-managerial role of *Homo sapiens* was advantageous. It helped enable our species to populate many parts of Earth's land surface.

Spatially separated human populations developed distinct cultures, diverse languages, various technologies, and differentiated modes of organization for collective action. Our species had first arisen in tropical surroundings, but as we grew more numerous some of us spread to places where cooler climates could have been devastating to non-fur-bearing bodies. We necessarily learned to use *prosthetic substitutes* for the fur we no longer inherited because our ancestors had adapted to the tropics. Skins taken from other animals could enclose portions of the human anatomy and retain metabolic warmth. Human feet that had evolved from the hand-like hind feet of quadrumana could be further improved for terrestrial walking and running by artificially converting them to more hoof-like form by enclosure in shoes, made from tough animal hides.

Clothing was only one aspect of dependence on prostheses. Little by little humans in various places learned to accomplish increasingly impressive things by using artificial devices that augmented their bodily apparatus in many ways. They derived energy from sources other than their own muscle power. From time to time equipment was devised that enabled the muscular-power of other animals to be harnessed for performance of human tasks. Eventually we also harnessed inanimate energy sources—flowing water and moving air. All these developments enhanced human ability to elaborate and diversify the structures and activities that made *Homo sapiens* so distinct from (and dominant over) the rest of the animal kingdom—making ever more plausible the idea that we had been divinely endowed with "dominion over the Earth and all its creatures."

———

It was not just *any* old hominid species that staked this claim to dominion. Most hominids, like any other animal, would have related to the rest of nature by finding the other types of creatures they encountered either predatory threats to be eluded, or prey to be exploited, or competitors. It was the one hominid type that had developed a very considerable tool kit, the one most committed to life shaped by a cultural heritage in addition to its genetic heritage—symbol-wielding *Homo sapiens*—that eventually undertook to manage local ecosystems.

Deliberate ecosystem management was a truly fundamental change, a real "breakthrough." The importance of having taken up farming, at first as a supplement to hunting and gathering, later as a virtual replacement for such a "pre-human" lifestyle, was that it enabled *Homo sapiens* to adjust upward the world's human carrying capacity to match our growing numbers—rather than having to accommodate our numbers to nature-determined carrying capacity limits. In other words, the dawn of agriculture was, like the much-later discovery of a second hemisphere, a significant raising of any Malthusian ceiling. Agrarian societies set about converting this planet indeed into "a world made for mankind." Human life flourished under conditions of a carrying capacity surplus contrived by human ingenuity.

Without quite realizing the real character of our breakthrough, we had become a "prosthetic species." Using diverse technological extensions of the bodily apparatus with which nature had endowed us, we began to behave as giants—with growing appetites for resources, increasing mobility, and enlarging quantities of end-products of our increasingly exosomatic metabolism requiring disposal.

After many centuries, without yet having acknowledged an ultimate global limit to the artificial enlargement of Earth's human carrying capacity, the technological extensions of our bodies upon which we had grown dependent came to include mechanical *muscles* (engines of various kinds). Unlike the muscles of draft animals which, like our own muscles, converted the chemical energy in foodstuffs into mechanical energy, engines were devised to extract energy from fuels rather than from foods. And it turned out there were vast stocks underground of combustible substances. The fact that those carbon-rich substances would burn, whatever other chemical and physical properties they might have, led us to define them as fuels. Not until centuries later would they be recognized as nature's way of having sequestered the carbon early organisms had removed from Earth's primordial atmosphere. That removal by those early species had fortuitously changed the world, giving it characteristics that enabled animals (including humans) to live and breathe, and endowing it with a temperature regime and hydrological cycle compatible with human needs.

Vegetation growing on present-day Earth continues to extract carbon from the atmosphere. The burning of some vegetable matter, the metabolizing of some of it in the bodies of humans and other animals, and the decaying of much of the total biomass in support of various decomposer microbes, puts carbon back into the atmosphere. Some vegetable matter escapes being eaten or burned or decayed and may end up being buried (sequestered). But the rate of carbon sequestration today by natural processes is minuscule compared to the rate at which prehistorically sequestered carboniferous materials are being exhumed and burned by industrial societies.

For more than two centuries increasing fractions of a growing population of *Homo sapiens* have been equipping themselves with more and more mechanical muscles—becoming *Homo colossus*, prosthetic giants with gigantic resource appetites, enormous mobility, and colossal disposal needs.

Fuel tanks of our machinery are the stomachs of *Homo colossus*. Filling them requires extracting from the underground deposits of coal, petroleum, and natural gas, prodigious quantities of that buried (formerly atmospheric) carbon. What we have called "progress" has entailed escalating demand for fossil energy. Our "energy supplying" corporations don't *produce* these fuels; they just unearth them. It was prehistoric organisms that produced them by photosynthesis, mostly millions of years ago. The stocks of buried carbon are finite. These "resources" are not renewable on a human time scale.

We have hardly begun to recognize how utterly different the Industrial Revolution breakthrough was from the Dawn of Agriculture breakthrough. Converting from the hunting-gathering way of life to the agrarian way was a shift from foraging to farming, from dependence entirely upon nature's largess to managing portions of nature in the exclusive interest of our own species. That revolution enabled us to live for a while under the happy condition of a carrying capacity surplus—a surplus that was increasing (but only temporarily). Converting from agrarian to industrial living—becoming *Homo colossus*—was a return to foraging. By defining those underground deposits of primordial atmospheric carbon as fuels, and committing ourselves to a way of living that required burning vast quantities of the stuff, without regard to the finiteness of these fossil energy stocks, we were undoing the raising of carrying capacity limits we had previously achieved.

Homo colossus is a forager species, seeking and consuming nonrenewable resources to satisfy gigantic prosthesis-based appetites. And we have been undoing—at an accelerating pace—the prehistoric conversion of Earth's atmosphere to the user-friendly form that made *us* possible.[263]

As we continued to multiply our numbers and enlarge our appetites for non-renewable resources, we swiftly converted carrying capacity surplus to carrying capacity deficit. As *Homo colossus* we became addicted to lifestyles that are unsustainable. *Homo colossus* is inherently a foraging species, dependent again upon limits imposed by nature, not repealable by human effort or ingenuity.

By means of the industrial revolution, converting ourselves from *Homo sapiens* to *Homo colossus*, we have undone the phenomenal achievement our ancestors made in raising human carrying capacity by the agrarian revolution. When our ancestors took up farming and were no longer foragers, they were making real progress. We, instead, have returned to the old practice in a big way and are hell-bent on foraging ourselves to oblivion.[264]

CHAPTER 12

How Money Led to Dehumanization

"Man exists as an end in itself. Therefore: 'Act in such a way that you use Man, in your own as well as in any other person, always as an end, never only as a means.' This is the supreme limiting condition of every man's freedom."

—Immanuel Kant, "Groundwork to a Metaphysic of Morals" (1785)

"There's nothing in the world so demoralizing as money."

—Sophocles, *Oedipus Rex* (428 BC)

"There's a sucker born every minute."

—P. T. Barnum

Division of labor has done away with self-sufficiency. Our dependence on a network of exchange among diverse specialist occupations involves money, so money comes to be a very necessary aspect of life. Has the obligatory pursuit of money forced us to violate Kant's maxim, so that we tend to act toward others as means only?

As we have noted in previous chapters, division of labor has been seen first as a way to achieve economic progress, but alternatively, as a means of social progress. For Adam Smith, specialization of tasks was the organizational mode that enabled workers to become more productive. The result was advancement economically toward product abundance. For Emile Durkheim, interdependence was what had to result from interaction of differently specialized workers, and—with or without ensuing abundance—enhanced interdependence was seen as socially beneficial because it would yield organic solidarity.

E. A. Ross, on the other hand, saw a lamentable aspect of the exchange process implicit in such interdependence—the emergence of new ways for humans to victimize other humans. To the "stealings . . . that lurk in the complexities of our social relations" he referred metaphorically as the *picking of pockets* by various industries' self-serving and often deceptive practices. We must now consider how apt is that pocket-picking metaphor, and how generalizable.

Use of people as tools

Division of labor spotlights a fact about people interacting with others that we don't commonly acknowledge. Although different specialists can be of mutual assistance to one another in the processes of living, the other face of that relationship is that each of us must *use* others—and be used by others. The garage mechanic who remedies a malfunction in my car's engine is useful to me. His tools are useful to him. To me, the tools in his hands together with his skill in putting them to appropriate use make him useful to me. In effect, his skill-and-tools become one entity, and *that entity* becomes my tool—the device I use to get the car running properly. In the navy in World War II, when I was an Aviation Machinist's Mate, I was a tool that helped keep certain aircraft airworthy for naval aviators to fly. We rely in everyday life upon many other human tools. They make things function the way we want them to function. When we buy a ticket and enter a cinema, our evening's pleasure involves using a projectionist, among others, to provide the screen images we paid to see. Quite probably we never see that projectionist, nor did we see the many people whose specialized tasks manufactured the photographic equipment and materials by which the performances of actors and actresses were recorded on film, available for projection on many screens.[265]

In the complex lives we live in modern societies we all *use* each other—to perform the myriad tasks we ourselves are not individually equipped or trained to do. We exchange our respective skills and services with one another—not directly in most instances, but indirectly through multiple buyings and sellings in a great web of trading. Each of us is a resource to others. Others are resources upon which each of us must rely. But every link in the web of exchange affords temptations to take shortcuts. Some shortcutting is considered decent and prudent—cutting out "the middle man" to minimize costs. In some cases, though, shortcutting amounts to cheating, or worse.

We had ancestors long ago who lived more simply. They too traded services and objects, but their trading web was very much more easily understood than ours in this present more populous and technologically advanced world. They lived without benefit of many things and opportunities to which we have become sufficiently accustomed that we consider them indispensable. Before there were switches on walls for turning on lights, nobody sat in a dark room resenting the inability to obtain illumination by so simple a finger motion, nor did those ancestors expect some remote manufacturer to provide such "magical" equipment. People lit candles. About the future existence of such electrical devices as incandescent bulbs and wall switches, or generators, transmission towers, circuit breakers, transformers, etc., people living in candlelit times were quite oblivious, and thus could scarcely even have wished for them. Their web of exchange was simpler, with fewer intermediate and beyond-the-horizon links. Even so, occasionally those ancestors living simpler lives may have exploited each other, taken some shortcut to possession of a coveted object (stealing, rather than purchasing, silver candlesticks) but their opportunities for shortcutting were limited just as were their kinds of activities and types of possessions.

From Shortcuts to Pickpockets

When *Homo sapiens* first came on the scene, there were fewer (and only much simpler) tools for their hands to wield, so there were fewer distinct tool roles for early members of our species. But there were some, and people, as soon as they were human, did use other people, in various ways. Human exploitation of humans goes way back in time. Slavery arose early on. Thievery also arose as soon as people had things other humans could forcibly or stealthily take from them. But there were no shoplifters until there were shops for selling things. Until the clothing humans wore had developed to a point where there were pockets to contain coveted artifacts worth snatching, there had been no pickpockets. Until people began obtaining some of their goods from afar, piracy and highway robbery were not yet possible. These criminal activities arose as plausible shortcuts to wealth when there were streams of goods that could be diverted from legitimate to illegitimate channels of trade.

When we use the term pickpocket literally we envision a person who extracts something of value from another person's pocket without the latter's consent (and presumably without the victim's immediate awareness). As the extracted "something of value" we probably visualize in most instances today a billfold or wallet—containing money. To pick another's pocket is to take a shortcut, acquiring money without the usual expenditure of labor or participation in trading something for something else.

Personally, I decided long ago never to carry my billfold in my back pocket. The back pants pocket is presumably the one from which a comparatively flat object could be most easily extracted without the wearer's awareness. There is no way of knowing for sure whether the habit of carrying my billfold in a less easily picked pocket is what has prevented losing it, but no such possession of mine has ever gone missing. There was one pickpocket by whom I was *almost* victimized during a vacation trip as I stood amid a crowd of tourists in the sanctuary of the Cathedral of Notre Dame in Paris. Others in my party saw her in the cathedral's semi-darkness lifting the edge of my coat and attempting to take hold of whatever might be in the back pocket of my trousers (it was *not* my billfold), and they and I saw her quickly darting empty-handed off into the crowd as I turned around.

When I was attending a professional conference once, in a city within the U.S. but nearly a thousand miles from my home, money *was* taken from my billfold while I was sleeping. The billfold was on the dresser in my hotel room, which the thief entered through the room's open window, having apparently come up the outside fire escape. He was a skilled and clever individual. Before exploring the contents of my billfold, he had quietly unlocked and opened the room's door to allow quick escape into the corridor, and it was the light shining in from the corridor that awakened me so that I saw him turn from the dresser and hastily depart. At first I feared he had taken the billfold itself, including my driver's license and credit cards—which would have left me facing dire difficulty in buying gasoline for the drive home. Happily for me, he took only my cash, to minimize his risk, if caught, of being found in possession of something quite visibly not his own.

That is an important characteristic of money. In its simplest forms money does not tell whose it is. Money can serve the person who has it, whether earned by its holder or obtained by theft. That fact underlies the pickpocket's "profession." In a modern economy, with extensive division of labor, we all use money (in one or another of its several forms). So we need it, and for most of us much of our activity is directed at acquiring it. Money buys the various resources we use. Most people are paid for their tool function—performing some form of work, or selling some product or commodity, sought by others (who have money to buy). How the various forms of work, or the various kinds of commodities are priced, can seem fair and equitable to some minds and grossly unfair and inequitable to others, but either way, they are part of the system of exchange that goes with being human in today's world.[266] However, since money is money, whether earned or not, by the exercise of whatever skill, there is incentive for people to help themselves to it when they can.

There is also an incentive to be thrifty. For anyone who has money, or expects to have it, there is a more or less constant urge to take measures to retain it, against whatever possibilities of its departing—either by superfluous expenditure, or through theft by others. In a money-using society (which is to say, any society with extensive division of labor), there is temptation in everyone to extract money from others, but by the same token there is a tendency for everyone to look upon everyone else as money-extractors—people seeking to extract money from one's own pocket or purse. When others regard us mainly as money-extractors it has to feel dehumanizing, just as when we know we are seen by others as mere tools or resources. People like being considered useful but resent feeling used.

From Pickpockets to Captains of Industry

People can be tempted to take shortcuts. Shoplifting and burglary are shortcuts, by which a person bypasses the usual steps in acquisition of desired items—earning a monetary income and exchanging some portion of it for the desired items. Picking a pocket is another mode of taking a shortcut, acquiring money more directly than by earning it. But shoplifting and picking pockets, or stealing in general, are not the only ways a shortcut to "wealth" can be taken. There are other, very lucrative ways—swindles, corruption, tax evasion, corporate malfeasance, etc. Ross chose his words to indicate that these shortcutting actions were like picking pockets on a grand scale. He wanted his readers to be more concerned with these larger sins against society than with petty crimes against individuals. My reason for invoking the pickpocket metaphor in this chapter is to call attention to the dehumanizing attitude arising among exploiters (on whatever scale) toward their actual or prospective victims, seen as mere bearers of billfolds.

Unless the billfold taken by a pickpocket is new, stylish, and expensive, it likely has little or no value itself to the pickpocket. It would be the probable *contents* of the billfold, some amount of currency, that was sought by the pickpocket's criminal act. A billfold is used by its owner as a "tool" for carrying money (and other items). A person with a billfold, then, becomes for the pickpocket a tool for bringing money into the proximity of his or her deft fingers.

Money, especially paper money, has virtually no *intrinsic* value. That amount of paper (even paper of banknote quality) is of little value until imprinted by government-prescribed design to represent a specified amount of "purchasing power." When the words printed on it declare it to be "legal tender" it becomes socially recognized currency. Then it may have sufficient value to be worth stealing. After the billfold has come into the hands of the pickpocket, whether it has yet been missed by its owner or not, the money in it represents purchasing power the pickpocket can use to obtain in transactions having the appearance of legitimacy any of various kinds of goods he or she may desire, just as the money would have been used by its rightful owner, now victimized and no longer in possession of it.[267]

For the pickpocket, picking pockets seems a better shortcut than shoplifting.[268] But when Ross used the pickpocket image *as metaphor*, he was concerned not just with that literal form of petty theft, but with the prevalence among powerful leaders in business and industry of a fundamentally similar shortcutting zeal. Rather than "making money the old fashioned way, by *earning* it*" through more or less honest exchange, people in positions of great economic power can obtain money in ways that provide less than a quid-pro-quo to its sources. People with a commercial outlook regard other humans not so much as ends but partly as mere means. Thus the common attitude toward others held by people in positions of economic leadership too often is fundamentally akin to the attitude of the pickpocket toward his target. So, in effect, Ross recognized the business baron as having a pickpocket-like mentality.

The business mentality he was metaphorically denoting is one that clearly violates that important ethical principle stated by Kant. The manufacturing and merchandising processes under the command of powerful men, and all the involved activities of a business enterprise, tend to be seen in essence by merchants and leaders of industry as *means of money-accumulation*. But to accumulate money means extracting money from other people. That view of commercial activity implies that *customers*, the people from whom money is to be extracted, are, in Kant's terms, not people as ends but people as means.

Roughly midway between Kant's enunciation of that ethical standard, and Ross's discernment of rampant departure from it, Karl Marx deplored (on a similar but limited basis) the relationship between capitalist owners of the "means of production" and their employees. When Marx argued that this relationship used labor as a commodity and creamed off the "surplus" produced by the workers, he was fixing his gaze too much upon only part of the dehumanization process that had begun to pervade industrial societies. Dehumanization characterized the attitude not just by factory owners toward the workers who were being treated as tools (means, not ends). Competing firms, suppliers of materials, customers, transport systems, *everyone*, when viewed as either revenue sources or as costs of doing business, tend to be considered in less than-human terms—as a consequence of the many-faceted interdependence induced by division of labor.

Of course to employees the employer ceases to seem fully human when regarded simply as a wage source and an exploiter of workers. But Marx was less concerned with workers' own violation of the Kantian principle.

Anyone with an accessible billfold can become the pickpocket's "mark." Anyone seen as a potential customer of an industry or business tends similarly to be regarded as just a source of "revenue." Anyone seen as a source of purchasable goods or services may be viewed by the potential customer less as a human being than as a vending machine (price of items vended thereby to be bargained down if possible). The implication of the pickpocket metaphor is perhaps even stronger than Ross intended when he used it. It implies that the business world's unacknowledged attitude becomes one which spreads through the whole population, so each of us tends to define other people basically as bipedal billfolds—from *which* (rather than from *whom*) money can be extracted.

Viewing "others" that way is just a slightly more abstract version of the perception of them as useful tools. Not that every member of a modern society sees *every* other member at all times as *nothing but* a walking billfold. But it is essential to delay no longer the recognition that *tendencies toward* such an attitude are nurtured by the experience of living in a society built on division of labor and the practices of exchange, necessarily using a monetary system. So I am here arguing that it is not solely the shortcutting captain of industry who sees workers as costs to be minimized as much as possible (and "laid off" when times are hard), and sees prospective customers as walking pockets from which to extract as much money as possible. Division of labor, which entails that whole web of exchange relationships, pressures *everyone* to respond to others less as complete ends-in-themselves human beings and increasingly as useful (specialized) tools—and/or billfold carriers. We are all being pushed by the nature of today's societies toward perception of the people outside our circle of face-to-face acquaintance as abstractions that are only vaguely human.

Linguistic Anesthesia

Sometimes we can learn important things about what goes on in people's minds by the kind of language they use. A few of the speech habits of pickpockets serve as indicators of their attitude about their "occupation" and about its victims. A century and a half ago the novelist Charles Dickens provided a lucid depiction of a gang of pickpockets trained and "employed" by an infamous master.[269] When the orphan boy, Oliver Twist, first came under the influence of this man Fagin, his initial exposure to the ways of thinking and acting characteristic of this criminal subculture was by means of a "game" in which pickpocket skills were practiced—the boys competing to pick Fagin's own pockets so deftly he would not feel it happening, even though as their teacher and leader he knew the game was on. When Oliver had a turn, after watching the others, and did successfully remove a handkerchief undetected from Fagin's pocket, he was called "a clever boy" by the playful old man, who added, "I never saw a sharper lad." In addition to such words of praise, Fagin further rewarded Oliver's performance by giving him a shilling, and added, "If you go on, in this way, you'll be the greatest man of the time." This left Oliver wondering how picking Fagin's pocket in play could have anything to do with achieving future greatness. He had not yet quite realized the group around him were serious pickpockets.

But later, when Oliver went out on London streets with Charley Bates and the Artful Dodger, heard their whispered designation of an old gentleman reading in front of a bookseller's stall as "A prime plant," and watched them take a wallet from the gentleman's pocket, Oliver fled in panic at the hue and cry and pursuit that began when the bookseller emerged and called out "Stop, thief!"

The word "plant" to designate a person to be victimized by members of this gang recurs later in the Dickens novel, and was apparently fixed in the lexicon of British 19th-century underworld activity. In the slang associated more recently with American pickpockets,[270] the intended victim is called a "mark," and the same term is used by con men to refer to the person whose confidence will be gained and then exploited. Just how fully such slang terms actually diminish in the mind of the pickpocket or con man the humanness of a targeted victim probably cannot be known. But it seems likely that a reason for using depersonalizing slang words is because they at least *somewhat* anesthetize whatever conscience the exploiter might otherwise have had.

Novelist Dickens used his fictional account of Fagin and the lads he employed as tools not to stigmatize such individuals but to illuminate and criticize the society that produced them and their subculture. It was the larger society's dysfunctional tendencies about which Ross was likewise incensed. The actions of "legitimate" business leaders which he so sternly likened to the picking of people's pockets, would generally involve much subtler, more indirect, and better disguised techniques of extraction. And whereas he sought to stigmatize such extractive procedures as monumentally criminal (even when legal), my purpose is less to stigmatize those procedures than to lament them. I simply mean to emphasize the *fact* that throughout all portions of today's complex societies people engage in activities that *are extractive*—aimed at extracting money from the billfolds of clients, customers, and taxpayers, or from the accounts of benevolent organizations, government agencies, etc. Whether or not any shortcuts are involved, any departure from the quid pro quo of fair exchange, routine acts by anyone which treat other human beings (and groups, even nations) as *revenue sources* are an inevitable aspect of the interdependence that has resulted from modern division of labor.

The temptation to treat others mainly as (or even as *nothing but*) revenue sources is omnipresent. In many minds and lives it may be repressed, disguised, even denied. But it derives from the need for money that is an essential instrument for the consummation of innumerable exchanges that became necessary when division of labor was elaborated beyond some threshold level. The temptation to perceive our conspecifics in dehumanizing terms is therefore pandemic. Its very ubiquity dulls recognition of it in ourselves.

Industrial-level division of labor was too new in Durkheim's time for this dysfunction of it to have yet risen to visibility. It was still possible in that final decade of the nineteenth century for his brilliant intellect to shrug off, as results from "abnormal forms" of the division of labor, the odd departures from the organic solidarity he expected. Not today. Instances of overt manifestation of the *nothing-but-walking-billfolds* attitude toward other human beings have become too frequent and too flagrant to overlook or explain away. Whether or not pickpockets in the literal sense as a category of petty criminals are on the increase, it is the pickpocket mentality in the minds and actions of "decent" people that has become too prevalent to disregard.

To observe the pickpocket mentality in action the reader need only refrain from pressing the mute button when the nightly TV news anchor speaks the magic word, "Next." (Whether we actually hold in our hand a real "clicker" or not, we all seem to develop in time an internal but only partially effective mute button somewhere in our overburdened brains.) That deceptive word "Next" will be followed, not by a specified news story, as implied, but by a series of commercials. Try *listening* to those commercials, and then imagine how they would have impressed Durkheim and Ross, respectively. If those two minds were to contemplate such sales messages *in such a context*, wouldn't both be appalled and dismayed?

Midas, Gambling, and Money Trees

Now it is time to consider other old-fashioned metaphors, representing other shortcuts. Magic was basically a package of shortcuts—ways of obtaining some desired result without going through all of the procedures actually required to bring it about. In our enlightened age most of us profess to have put behind us any belief in magic, but we sometimes fantasize about what we would do if given a magic wand, and our enjoyment of literature about persons with magical powers is not always confined to our childhood years. Even as sober adults we retain some very magic-like expectations in connection with the world of money.

We may scorn alchemy, but we still remember the legend of King Midas. He had the magic touch. He wanted abundant gold, and his wish that everything he touched might turn to gold was granted. Then, rather than having to find a location where the rocks contained veins of this precious metal and undertake arduous mining operations, he just touched things and they became gold. Why did he want gold? Already in that legendary time, gold was a measure of wealth and a medium of exchange, so possessing gold was to possess the power to purchase all manner of desiderata—even in an era of much less elaborate division of labor than engulfs us now. So the imaginary Midas, by this magic, had availed himself of the ultimate shortcut.

It is not certain that modern people would shun an opportunity to partake of his wonderful power, or some apparent approximation of it. People who visit a casino or buy a lottery ticket or enter a sweepstakes may persuade themselves that they are engaged in a form of recreation, and the casino industry encourages that self-deception by avoiding the term *gambling* and insistently using the word "gaming" instead. But wherever they are legal these businesses flourish because seekers of shortcuts-to-wealth still have the urge of Midas. People are attracted by free gold, gold with magical purchasing power, gold they didn't have to work hard to excavate from the Earth. Money in one form or another is desired for its power to purchase many other things, things we'd like to obtain without effort, perhaps in an abundance surpassing a lifetime's true earning power by honest labor.

But alas, Midas found this magic a very mixed blessing when a beloved daughter whom Midas fondly embraced turned instantly into a solid gold statue! The ultimate dehumanizing repercussion of the ultimate magical shortcut!

For how many big winners in state lotteries is there a similarly disheartening eventual outcome of unearned wealth, some unforeseen tragic disruption of their lives? Lack of foresight about the full range of consequences that would be wrought by the magical power he sought led to personal disaster for King Midas, and that was the intended moral of the legend. Why do the news media give us glowing stories of suddenly wealthy lotto winners on the day they get their "good news," and seldom follow up a year or so later with stories about their changed lives? Why are there no reports of the vast numbers of *losing* ticket-buyers? The power of gold can impoverish as well as enrich.[271]

What does the ancient story of Prometheus, a Titan (cross between a god and a human) and regarded as a benefactor of mankind, tell us today? Even though his name derives from the Greek word for foresight, did Prometheus adequately consider the harm that could be done by the human species which he equipped with the power of fire, stolen from Zeus, the CEO of the Olympian gods?[272] Acquiring the ability to use fire was an important step in the evolution of the human species. But when our fire-using ancestors defined as fuels the buried substances from the carboniferous age, rather than perceiving them as safely *sequestered* atmospheric carbon, they gave themselves and their descendants, by this Promethean definition of coal and petroleum, a tragic Midas touch—the power to wreak havoc with the planet upon which we must depend.

Avoiding Monte Carlo and Reno and Atlantic City, and leaving Midas and Prometheus confined in story books to bemoan unforeseen outcomes, consider an old metaphoric cliché. It may or may not be as familiar today as it was when I was a child. Think of the proverbial tree on which money does not grow. At least during the economic depression in the 1930s, parents who not surprisingly had urgent need to teach standards of thrift to their offspring, commonly voiced the warning, "Money doesn't grow on trees," meaning that money is harder to obtain than the leaves and fruit that grow on trees. Insofar as that cliché instilled the intended "respect for the value of money" in the children who heard it, it served as code for "you have to work to earn the price" of what you desire. A child's weekly allowance in those years had to accumulate for many months before it sufficed to buy some of the things children could even then easily learn to want. Probably a good many boys and girls therefore occasionally had wishful reveries about finding a money tree growing in their back yard.

Ironically, though, there are numerous examples of mature, educated people so behaving that we can infer they regard the world as an orchard of money trees. The "work" they do amounts to contriving ways to harvest the pecuniary fruit of those "trees" as easily as their ingenuity will permit. As Iacocca told us, a role requirement for the salesman is to keep those products of industry "moving."

What Kinds of Trees Bear Such Fruit?

There are many organizations and institutions that people treat as money trees. Occasions arise when officials of a corporation may begin to treat the company not just as an organization to be administered in ways that serve *its* reason for being, but as a money tree to be plucked for *these officials' personal benefit.* Governments, on the other hand, may be so regarded by many people, in many circumstances. And there are glaring instances in which the consumer, the public, the masses, are so regarded. In addition, there are cases of money tree plucking, combining aspects of all three.

1. Corporate money trees

The fate of the ENRON Corporation was a flagrant example of a corporation whose own officials used it as a money tree. Whatever "fruit" such an energy-trading firm might have ever produced, its very "foliage" seems to have been rapaciously harvested by its own leading officials.[273] Officials high in the corporation hierarchy apparently were able to maximize their personal wealth by manipulating company financial records in ways that kept the price of company stock rising, enabling their "stock options" to escalate in value, even while their corporation was heading toward bankruptcy.

And the Enron case was not unique. Between 1995 and 2005, the average annual compensation received by CEOs of 350 major companies rose from $2.7 million per individual to $6.8 million, a 151 percent increase, but in this period these companies' median sales rose only 51 percent. The CEOs' pay included their bonuses and stock options as well as their salaries. In the 1970s most CEO pay had been in cash; by 2000 half or more tended to take the form of stock options.

The phrase "stock options" refers to a corporation official's option to take some of his compensation as shares of his company's stock (reckoned at a fixed price, regardless of whether actually taken then or bought later), and then sell those shares for whatever their market price has later become. Ostensibly this was to motivate company officials to do all they could to maximize company efficiency and profitability. But it too often (in other corporations besides Enron) led the officials to manipulate reported profits (not always honestly) so as to enhance the market value of company stock they either held or could acquire at that prior fixed price (and then sell at the newer, higher price). They took a shortcut to wealth. The effect was to enrich the officials of the company at other shareholders' expense—and at dire cost to ordinary employees if and when the corporation went broke.

In the present context what is significant is that these manipulative officials who unduly enriched themselves and did grave financial damage to others seem to have contrived a moral code exempting themselves from normal constraints.[274] In short, they behaved (on a grand scale) as if the many others—investors and workers adversely affected by their actions—were nothing but pockets containing billfolds from which they were entitled to extract money by whatever means they could devise. It was as if the corporation they ran were a money tree which their positions enabled them to strip of its golden foliage.

The Enron example was by no means unique. In July 2005, another company's CEO, one of the most prominent in the United States was sentenced to 25 years in prison for securities fraud, false statements and conspiracy. He had been convicted on charges arising from a scheme to inflate the earnings and understate the expenses of the corporation he headed—by an $11 billion total. In addition to facing the prison sentence, the *Washington Post* reported he had had to give up his mansion and other assets worth a possible $45 million to settle a class action suit by investors who were defrauded.[275]

Examples such as these are cited here not just to take note of frauds, as such. Fraud, presumably illegal in most instances, would qualify as one of Durkheim's "abnormal forms" of division of labor. So not just fraud, but *routine injustice*, is what we must learn to see in these cases cited to emphasize how my argument goes beyond Durkheim's acknowledgment that there could occasionally be "abnormal forms." It is important to recognize the fact that high and mighty people in a modern complex society, people who regard themselves as respectable leaders, who may have been well-esteemed by the public and the media in their home cities, can fall into a pattern of acting in the pickpocket way toward many, many other people. They have in fact treated investors, as well as diverse categories of workers in their companies—and nameless, unseen *customers*—as pickable pockets. Their lives—and the lives of many of those workers and those investors too—are lived by money-extraction standards. My point is this: Money-extraction standards for relating to members of our society are an expectable consequence of our highly ramified division of labor. They are a significant aspect of the interdependence that was supposed to provide a modern society with organic solidarity. Societies seem not yet to have devised adequate means of curbing the manifestly dehumanizing influence of the money-extraction norm.

There was a time when CEOs of big corporations were regarded as a superior class of human being. They were assumed to have a multiplicity of skills essential to the leadership roles they played. They had to spend time in a number of places, and had to keep in their heads huge bundles of information about their company's activities. Because it was assumed no ordinary person could perform the CEOs' elaborate tasks, it was supposed their enormous salaries were justified compensation.

Recently, though, some CEOs have tried to assume a very different image. They have tried to persuade juries (and the American public) that they were "hapless dupes." When the WorldCom CEO, was shown that his company had hidden billions of dollars in expenses, he claimed to be totally ignorant about any deceptive accounting. The same sort of claim of innocence by ignorance has been voiced by, or on behalf of, other corporations' leaders in other cases.

In each instance, if the company's shareholders were on a sinking ship, so was their ignorantly innocent CEO, or so it was claimed. He or she purported to have simply trusted the accountants' quarterly reports. It was just some rogue underlings who had misallocated funds—even, allegedly, the person who occupied the office next door to the CEO's own splendid office.

More than one firm's CEO, whose company was headed into bankruptcy, or has been sued for fraud, has portrayed himself as having no mean streak, claiming to be just an innocent and astonished victim of corrupt or treacherous employees.[276] The first instance or two of this kind seemed like some sort of historical anomaly—as if the CEO who didn't really understand what his own company was doing had to be a rarity. Now that a number of trial defenses have pleaded innocence by ignorance, though, and such a management style has come to seem to be the norm, a larger inference has to be considered. Is being a CEO of a giant firm, whatever else it may be, also a special opportunity for picking the pockets of investors and workers? Do the perquisites of power foster an uncritical sense of entitlement to the contents of all those other people's billfolds out there?

2. Governments as money-trees

Corporations in danger of going broke sometimes turn to governments for bailouts. The bailout may take the form of a direct grant, an actual subsidy of the faltering company, or a contract for government purchase of the company's product which seems somehow to give greater benefit to the seller than to the purchaser—a sale in the company's interest more than in the government's (the public's) interest. Serious loss of "market share" by a corporation that trades internationally may trigger the effort to get help from government, as conditions contributing somehow to foreign competitors' increasing sales, or domestic producers' declining sales, are felt to be "unfair" and in need of "correction."

Industrialists' requests for help from government may be euphemistically stated as expressions of desire for legislation or administrative regulation to "level the playing field," based on allegations of something "unfair" about the competitive situation. For example, the image of a playing field needing to be leveled was invoked when the Ford Motor Company's new chief executive, Alan Mullaly, formerly in the aircraft industry with Boeing, planned to talk with the U.S. President. In the first 9 months of 2006 Ford had lost $7.24 billion, and was planning to close down a number of its plants, and was offering early retirement packages to all 75,000 of its U.S. production workers.[277] General Motors had persuaded about 35,000 hourly workers to leave the company under early retirement or buyout plans.[278] The American Big 3 automakers were seeking amelioration of problems arising, in their view, from such things as currency exchange rates and health care costs. From their perspective, huge financial losses were due at least in part to difficulties in the international market.

The auto executives seemed oblivious to the possibility that an era was ending—a time of cheap and abundant fuels, during which daily movement of Americans (and to a considerable extent people in other lands as well) across considerable distances was primarily by private automobile. To them it was unthinkable that this means of mobility—so long deemed normal and unchallengeable—could be rendered obsolete by changed circumstances. These executives were adamant that they were only seeking federal government effort to *level that playing field*, "not a government bailout." But the situation and the discourse were reminiscent of earlier "rescues" of corporations by federal actions—most notably the loan guarantees provided to Chrysler in 1979 that kept it from having at that time to declare bankruptcy. At that time cartoonist Don Wright of the Miami News drew a parent and child standing in front of the U.S. Capitol building, with the child asking, "What do they call that place, anyway?" He had the parent replying, "The Chrysler Building!"[279]

In 1979, the auto industry executives attributed their woes to "overregulation" by government, and when the head of Chrysler (who had formerly headed Ford) went to seek help from Congress, he argued it was "only fair" that insofar as government had contributed to the industry's troubles it should contribute to the remedy.[280] And he enumerated various instances of previous federal aid to beleaguered industries, and other entities. The two best known precedents were guaranteed loans to the City of New York in 1975 to save it from going bankrupt, and in 1971 $250 million in federally guaranteed loans to Lockheed Aircraft to save its workers and suppliers. A loan-guarantee board was established by Congress to oversee that operation. Lockheed did repay its loans, plus $31 million in fees.

Government loan guarantees seemed to Iacocca "as American as apple pie." They had been granted to "electric companies, farmers, railroads, chemical companies, shipbuilders, small-businessmen of every description, college students, and airlines." There were, he said, "$409 billion in loans and loan guarantees . . . outstanding when we made our $1 billion request," even though most people at the time thought loan guarantees to Chrysler would be setting a new and dangerous precedent. But five steel companies had received loan guarantees under the Import Relief Act of 1974. Iacocca also cited loans to the housing industry, subsidies for tobacco farmers, and loans for airlines.[281]

By a vote of 271 to 136 in the House and 53 to 44 in the Senate he got a package of loan guarantees (with strings attached) that did, for the time being, keep Chrysler going as the smallest of the American Big 3 auto making companies. Whether this rescue was going to be permanent or only temporary remained to be seen.[282]

Some industrial firms have come to regard the military establishment as a major money tree. Representatives in Congress from districts where such firms are located commonly seek credit for helping to arrange lucrative contracts. Local communities whose economies depend on nearby military bases commonly resist when base closures become likely.[283]

But there are other money-tree-plucking ways in which cities regard the federal government's departments and bureaus. Take an old example—San Francisco capitalizing on widespread sympathy after the devastation it had suffered in the 1906 earthquake and fire, obtained permission from Congress and President Wilson in 1913 to develop a new city water supply by damming a valley known as Hetch Hetchy *within Yosemite National Park*. The backlash against this economizing desecration of a wilderness area within a major National Park has swollen in recent years to a level that prompted a California state government sponsored feasibility study regarding removal of the O'Shaughnessy dam to "restore" that once pristine valley. Such a project was estimated to cost between $3 billion and $10 billion. It could take a century or more after dam removal for the muddy valley floor to return to being something like a lovely wilderness meadow it once was.[284] These estimates serve as a rough measure of the magnitude of the plucking San Francisco did almost a century ago to a Yosemite bough of a national money tree.

3. The people as a money tree

Whatever the medium—newspapers, magazines, radio, television, cinema screens, billboards—most of the advertising "messages" presented therein represent the use of language and other symbolic materials as tools for the extraction of money from an audience. Ads supposedly sell, or "move products." They are a manifestation of the mercenary standard for human interaction. Their existence implies that the people who are expected to be influenced by them are viewed by the merchant and the ad agency as pockets to be picked, although they may be more euphemistically referred to as customers, or even more euphemistically as "clients."

In the terms suggested by Emerson's power-dependence theory (cited in Chapter 8), an advertising industry is an enterprise devoted to inducing unbalanced human relations—making consumers more dependent, and sellers more powerful. Ads are devised to foster acquisitiveness, to persuade people to be dissatisfied with whatever items they currently possess, with experiences they have already had, and even with who or what their present self-conception says they are—so they will buy the new product or the new service.

Here is an indication of how far this can be carried in a modern society. In July, 2007, *Parade Magazine*, one of the syndicated supplements inserted into various Sunday newspapers across the United States, asked readers to come up with innovative solutions that might help reduce the need for the U.S. Postal Service to keep raising the postage rates. One writer submitted the proposal for the postal service to obtain revenue by selling advertising on its stamps, thus keeping down the cost to letter-writing consumers. The so-called "idea gurus," who had written the piece soliciting ideas from readers, said they liked that one. Magazines, and broadcasting stations, "are supported by ads," they noted. That being so, they asked rhetorically why shouldn't the post office get in on the same practice?

Clearly, their concept of "costs" meant *monetary* costs. But there may be non-monetary costs in addition to the price of a stamp. The post office does issue "commemorative" stamps, which have a function additional to obtaining money in exchange for carrying the mail. They remind stamp-buyers of historical persons, places, or events—and awareness of such elements of a national heritage may be considered an aspect of good citizenship. The postal service thus *serves* citizens in more ways than one. However, if a postage stamp contains advertising, this implicitly defines the post office as, at least in part, a merchandizing agency, existing to serve the advertiser, rather than an agency of the people, providing mail delivery service to facilitate human beings' communication with one another. Degrading the postage stamp by making it a medium of advertising degrades the postal service, which implicitly degrades postal customers.[285]

In recent years Americans have become conscious of, and irritated by, their country's dependence on oil from unfriendly sources, without generally seeing the situation in Emerson's power-dependence terms. American dependence upon Middle Eastern nations gives those nations power to affect American policies. Likewise, individual American motorists' dependence on suppliers of motor-fuel gives those suppliers power over motorists' pocketbooks. When gasoline prices have risen, many people have been predisposed to feel that the major oil companies were "gouging" their customers, rather than seeing how the rising cost of driving cars could be symptomatic of the finiteness of Earth's petroleum deposits. Congressmen have conducted well-publicized "investigations" of oil company policies, and eventually the companies are usually "found" not to have indulged in unwarranted "price-rigging." Public suspicion persists, however, and its existence should be seen as an important indicator of people's resentment over their aggravated dependence as consumers on powerful corporations.

In the summer of 2007, a document I encountered prompted me to write to the Chief Executive of Britain's Vodafone Group Plc, about his choice of words for describing a "strategy" designed to take advantage of "new technologies." His word choice had struck me as acutely symptomatic of the dehumanizing pickpocket mentality. Before using his words as an example in this chapter, however, I wanted to be sure the company had meant the phrasing as it sounded. I doubted my inquiry would receive a reply, but here is what I wrote:

> In your 12 June letter to shareholders, under the heading "Revenue stimulation and cost reduction in Europe," you said, "In Europe our focus is to drive additional usage and revenue from core mobile voice and messaging services and to reduce our cost base."
>
> I am wondering if there was a typographical error. Should the verb "drive" have been "derive"? Please enlighten me on this.
>
> It seems to me the difference in tone between these two words could be taken as indicative of a considerable difference in corporate attitude toward customers in the European market.

I had thought the implied dehumanization would have been appreciably milder if the company sought to "derive" additional revenue by means of additional usage of its product than if it really intended to *drive* additional usage and revenue. My letter was about a difference between two verbs. To my surprise, the following month a very succinct reply did come. I was not surprised that it came not from the Chief Executive to whom I'd written, but from a "Deputy Group Company Secretary." But I *was* surprised to find that he construed my inquiry as a question about a possible difference between two markets, not between those two verbs. Here is what he wrote:

Dear Sir

Thank you for your letter dated July 9th. I believe the word 'drive' was deliberately used and was intended to convey a sense of impetus in the stimulation of revenue from core mobile voice and messaging services. I don't believe the term was intended to imply a difference in corporate attitude toward customers in one market over another but rather as a reflection of the different market conditions we face in the regions where we operate.

Thank you for your interest in the Company.

Sincerely yours . . .

Written on behalf of a legitimate telecommunications company, not a fraudulent Enron, this letter and the policy it states could hardly have been dismissed by Durkheim as resulting from an "abnormal form" of the division of labor. The second sentence in this reply would seem to confirm existence of a company attitude that virtually epitomizes the view of customers as a money tree with pluckable foliage, or as a population of pickable pockets. That "disclaimer" sentence seems intended to assure me that any "difference in attitude toward customers in the European market" (as contrasted with customers in some other market) merely takes account of "different market conditions" in different regions. It essentially reaffirms the pickpocket outlook. It says, in effect, wherever they may be located, customers constitute a mere *market*, from which to extract revenue, pursuant to the company "impetus." In short, because the money-extraction standard is so deeply entrenched, the writer of the reply was deaf to the meaning of my comment about a "difference in tone" between two verbs—"drive" and "derive."

Tobacco Industry Dehumanization

There is perhaps no industry in which the pocket-picking attitude and the money-extraction norm has more emphatically superseded any semblance of the golden rule than the one which caused former U.S. Surgeon General C. Everett Koop to express deep dismay at public complacency about nicotine addiction. Despite the long-established adverse health effects of cigarette smoking, including more than a third of all cancer deaths, and many more from heart disease, millions of people continue buying and smoking cigarettes. Many smokers become afflicted with arteriosclerosis, peptic ulcer, and pregnancy complications, etc., and emphysema occurs almost exclusively in smokers. The public would demand action from Congress, Koop wrote, if any substance other than tobacco "were killing half a million people a year."[286]

Any doubt that the pickpocket attitude dominates modern society should be dispelled by the image of those tobacco company CEOs insisting before an investigative panel in Washington that they did not believe tobacco was addictive—despite the industry's deliberate manipulation of nicotine content to ensure continued sales of cigarettes to unable-to-quit smokers.[287]

Companies should not be allowed to make a profit from a clearly lethal product, suggested a former head of the Food and Drug Administration.[288] Society has tolerated tobacco companies shaping public perceptions of cigarettes and promoting cigarette addiction. The FDA uncovered clear evidence of deception and of nicotine manipulation, but the industry continued to argue that because tobacco is a legal product it should be treated like any other legal product, even though tobacco products kill people *when used as intended.*

The FDA long sought to make it harder for children to buy tobacco products, and it wanted to make such products appeal less to possible purchasers. Efforts along those lines made sense within the limits of FDA jurisdiction. But the companies' compulsion to increase revenue always trumped all such considerations and they somehow managed to sidestep every measure designed to reduce sales of their products. Not until some agency was empowered to regulate the purveyors of tobacco products rather than just monitoring the technical qualities of those products, could this evasion be expected to end.

Lawyers for the tobacco companies worked hard to prevent defeat in class-action suits. They followed the industry's public relations strategy which claimed the tobacco companies had already reformed. On July 14, 2000, a Florida jury awarded punitive damages of $144.8 billion. Tobacco products had killed 12 million Americans since the famous 1964 Surgeon General's report declaring them a serious health hazard. But cigarettes continued to be sold after this subsequent Florida verdict, and with each passing day, three thousand American children began to smoke. Despite trying to sound conciliatory, the industry did not stop fighting regulation, implying the verdict might not withstand an appeal.

Attitudes of tobacco investors were little affected by verdicts against the companies. Attorneys general in states that sued the tobacco companies publicly indicated they had no intention of pushing them into bankruptcy. As this showed, money was the big issue for the states which had sued. Their top priority after successful litigation was to ensure settlement funds would flow into state treasuries. The billions they were to receive over twenty years was such a tempting windfall that state legislators became intimidated about whatever might risk drying up the anticipated stream of cash.

"My understanding of the industry's power finally forced me," wrote David Kessler following his FDA experience, "to see that, *in the long term, the solution to the smoking problem rests with the bottom line, prohibiting the tobacco companies from continuing to profit from the sale of a deadly, addictive drug.* These profits are inevitably used to promote that same addictive product and to generate more sales. If public health is to be the centerpiece of tobacco control—if our goal is to halt this manmade epidemic—the tobacco industry, as currently configured, needs to be dismantled."[289]

The Restoration Dilemma

If *all* modern pickpocket-motivated industries were dismantled, what would be left of the human societies in which we actually live and upon which we have come to depend? It would amount to dismantling the modern system of division of labor.

The overwhelming effect of that would almost certainly be equivalent to the extinction of *Homo colossus*, whose colossal existence has depended on the many advantages yielded by the vast panoply of specialized occupations evolved over the recent centuries. Those diverse specialties are coordinated by exchange networks mediated by money, which inevitably aroused Midas-like aspirations among people whose individual self-sufficiency has been extinguished.

Can a society embracing anew the golden rule possibly become our future? Can humanity ever learn to abide by the neglected Kantian principle requiring humans to be treated as more than tools? To ask such questions is to wonder how well (and how willingly) could we revert from our *Homo colossus* dead-end lifestyle to being just a wiser, more modest version of *Homo sapiens*.

CHAPTER 13

How Obsolete Word-Maps Need Updating

"There is one thing that is worse than ignorance, and that is to know a lot of things that aren't so; or, what amounts to the same thing, to have exaggerated notions of the importance of what we know just because we know it, or because such knowledge has been the perquisite of the socially privileged in the past."

—George A. Lundberg, *Can Science Save Us?* 1961, p. 84

"Typically, *arrogance* is considered to be the antonym of *humility*, but I believe arrogance is, rather, a symptom of humility's true opposite—forgetfulness . . . . Refuge in our knowledge is perhaps our favored species of forgetfulness."

—Lyanda Lynn Haupt, *Pilgrim on the Great Bird Continent*, 2006, pp. 108-109

Languages evolve. People's vocabularies change as time passes. Some of the changes matter greatly and others may be relatively inconsequential. In the early days of the automobile era, people spoke of going places in someone's "motor." Later it was someone's "auto." Mostly nowadays it would be someone's *car.* Those verbal changes hardly make much difference in the way people live and behave, how possession of these transportation devices may be regulated and taxed, nor in the consequences of using these vehicles.

Whatever you call it, your motor/auto/car serves to carry people from place to place, its use contributes to the congestion of city streets and highways, and it burns fossil fuels and emits greenhouse gases. It will cost you substantially to insure it, to maintain it, and to use it. Changing the nomenclature doesn't change those facts, nor would they be different if the vehicle's common name had never changed.

Amid change there are cultural lags in our uses of words (and other symbols). Some language habits outlast the circumstances that produce them. Their misfit may or may not cause trouble.

Anachronistic verbal habits *may* merely indicate lack of fashion concern, but sometimes they can impede adaptation to changed circumstances. When *Homo sapiens* was a much less numerous species and when our ability to exploit the components of the Earth and products of the biosphere upon which our lives depend was much less developed than today, word-maps arose that have outlived their former validity. They now obstruct recognition of the way we have painted ourselves into a corner. Mankind has an urgent need to grasp the serious discrepancy between certain obsolete language habits and the true characteristics of the situation they misrepresent.[290]

Our use of language is a fundamental fact about our species. In part this is because it gives us a more intricate capacity than any other species has for sharing the facilities of one another's sense organs and brains—and for evolving and using a cultural heritage.[291] The human capacity for culture is what has enabled us to elaborate our *within*-species division of labor (and thus our power to transform the world) beyond anything achieved by any other single species—and almost beyond the most complex *inter*-species webs of symbiosis observed in nature.

As we have seen, this capacity for culture and the resulting (and ever increasing) quasi-speciation (division of labor) has been very useful to humankind, but it has had, and is having, serious "side-effects"—unanticipated consequences, some of which may be quite harmful to people, or to the environmental systems upon which people inevitably depend.

Overcompensation

But unfortunately an odd form of social pathology can arise, instead of wisdom, from recognizing the cultural lag in language habits. The causes, consequences, and varying incidence of that pathology merit serious scrutiny. Too often, recognition of the malfunctioning of verbal habits is pursued with a cult-like interest. When done in that unsophisticated way there often develops a too-supercilious attitude toward language, giving rise to the notion that word maps are *inherently* dysfunctional, that all words always deceive.

In an age of increasingly pressing ecological constraints, humanity is ill served by supposing that the fallibility of word-maps implies that any choice among them is purely and inevitably arbitrary (and that the "territory" we map verbally is therefore either totally malleable or quite fictitious). Yet that solipsistic sort of philosophy began to seem alarmingly fashionable among university students in the 1970s. It was one expression, perhaps, of an antiintellectualism that was fostered partly by the social disillusionment of the Vietnam War years[292] and, more subtly, by the discovery of facts we are reluctant to face about man's relation to the biosphere. That reluctance is still with us, so the folly of considering all statements of fact indistinguishable from fiction continues to plague society.

If we could not be human without language, then we need to *limit* our suspicions about words. We need standards for *reasonable* caution about the effects of language. We need guidelines for establishing and maintaining confidence in the affirmative utility of *valid* word-maps.

When we come to grief from following an obsolete word-map, we need to remind ourselves that over-generalizing is irrational. It is more rational to seek an updated word-map than just to deplore word-mapping as an apparently futile practice. Without word-maps we would revert to a more chimp-like status. We could not be human. When we say language habits can become obsolete, we should not seem to be saying language always misleads. If some words are meaningless and we sometimes mislead ourselves by faulty language habits, we should not conclude that words and word-maps are never useful and always mislead.[293]

Nature's Dictionary

Obsolescence, we must remember, is a relation, and relations can exist only between two (or more) entities. Thus a word-map can be obsolete only in relation to some specifiable territory having specifiable features. Both sides of the relation are considered in the following suggestive statement:[294]

> Words like *limitless, inexhaustible*, and *boundless* have figured prominently in the debate about the Earth and its resources. They have been persistently used despite the fact that they would not exist in a "Dictionary of Nature."

It is as important to study the changed circumstances constituting the territory as it is to study the social history of a word-map that has lost the correspondence it had with a former reality. Western man formed certain language habits during the centuries of expansion into a "New World," which was a wondrous environment whose carrying capacity at first so much exceeded the settlers' originally small numbers that it was plausible to suppose words like *infinite* were applicable.

The world's carrying capacity no longer exceeds our greatly increased numbers[295] and the centuries of migration into the western hemisphere have also seen at least a third of Earth's human population technologically transformed into *Homo colossus*. So continued use of such cornucopian words (and persistence of such thought ways) now have no justification. Because they persist nevertheless, Americans of good will, following word-maps based on such habits,[296] have often advocated an American standard of living, or something approaching it, for the entire world. "Freedom from want" was a goal held before the less prosperous peoples to enlist their support during the Second World War. It was a noble dream but a monstrous deception of ourselves and others—and as it turns out a crime against posterity. This should be clear now to anyone who at long last has come to think in terms of the carrying capacities of the lands supporting swollen human loads.

Our too-conventional word-maps still portray "developed countries" like the United States, major European nations, and modern Japan as the model for the supposed future condition of today's "underdeveloped countries." For most people it has been unthinkable that instead the trend might someday go in the other direction—that the prosperous industrial nations could in future decades descend to conditions more akin to the past and present poverty of the "undeveloped" or "underdeveloped" countries. Public comprehension of the reasons why this might be so has barely begun.

The new word-maps needed for living with the new circumstances must give prominence to the not yet widely familiar concept, "carrying capacity." That is a truly pivotal term in the dictionary of nature. The finiteness of the world must be recognized, as well as the inverse relation that ultimately holds between standard of living and numbers of users of a finite habitat with finite resources. The more progress made in the near term by what we used to speak of as "backward countries" and more recently had learned to call "developing countries"—the more they "catch up" to the "developed" countries while Earth's exhaustible resources last—the more severe will be the catastrophe when a greater-than-ever carrying capacity *deficit* takes its toll, as it ultimately must. Saying so is to state a fact that must be faced. It is not just a rationalization of past imperialism nor merely an ethnocentric disparagement of non-European people. The longer the developed countries hold fast to their imprudent planet altering ways the more abrupt and catastrophic the crash that will befall them *and the developing countries* as the false promise of universal development is refuted by nature.

Facts of Life Inverted

When settlement of the New World by Europeans began, the fundamental fact of life for the European members of species *Homo sapiens* was that their newly enlarged environment had a potential carrying capacity greatly exceeding their numbers. This fact shaped their outlook on life, making perpetual progress toward individual liberty and ever-higher standards of living seem not only possible but virtually inevitable.

Now, however, the old New World is more densely populated than Europe was at the time the great migration began. Furthermore each of the present inhabitants of Europe and America has more mobility and gadgetry than their forebears. Thus he or she uses far more of the planet's natural resources in their prolonged lifetime than did the average shorter-living and more modestly equipped inhabitant of sixteenth-century Europe.

Today's most fundamental fact of life is that our world's carrying capacity is not unlimited. In our daily lives we increasingly feel the effects of the limits, though too many people still do not understand that this is what is happening to us.

Moreover, in the first century or two after Columbus, even European *Homo sapiens* was still living very largely by the use of renewable resources. Recently, however, human activities had come to depend upon the use of some ten times as much energy derived from fossil fuels as people on average derive from organic sources or from moving air and moving water. An updated word map would thus have to include the information that only about one-tenth of the energy base for modern human activities comes from renewable resources.

The updated word-map should make clear that a "developed" country is one that has staked its future on continuing use of exhaustible resources. An "underdeveloped" country is one that has not so far been able to make that prodigal commitment.

President Roosevelt once said, "Our earth is only a little star twinkling in the universe—yet we can make of this—if we care to—a planet undisturbed by wars, unperturbed by want or fear." No American president since the Second World War has escaped the delusion that results from ignoring the concepts in nature's dictionary. All have failed to comprehend the basic change in humanity's ecological situation.

President Truman, through his Point Four program, expressed a similar faith to FDR's. President Eisenhower, visiting New Delhi in 1959, told throngs there: "We have today the scientific capacity to abolish from the world at least this one evil. We can eliminate hunger . . ." President Kennedy sought to revitalize traditional optimism in the United States with his concept of a "New Frontier," and for the whole Western hemisphere by launching an "Alliance for

Progress." After Kennedy's life ended suddenly in Dallas, President Johnson echoed his predecessor by exhorting Americans (as we recoiled from the shock of presidential assassination), "Let us continue!" His administration envisioned imminent achievement of a "Great Society."

President Nixon used much the same sort of rhetoric. To focus on a specific challenge, he attributed "the energy problem" not to an actual depletion of resources but to the fact that "the people of the world . . . are living better." Improvement of living standards, he said in a commencement address in June 1973 at Florida Technological University, had produced a "temporary problem," but long-term, he said, we had "an opportunity to fill the demands of all the people of the world."

This belief that in the long run the ever growing demands of all the people of the world can be filled was reiterated or implicitly assumed by each of the subsequent U.S. presidents, but it is a relic from an era that no longer exists—the era when, by discovery of land in a second hemisphere, available human carrying capacity had suddenly been raised far above the magnitude of the load imposed by the human population then existing. Access to a New World not yet filled up with voracious resource-users was the basis for a magnificent carrying capacity surplus. During more recent administrations, the obsolescence of that cornucopian belief became more incontrovertible but remained unacknowledged by most politicians.

Ordinary people cling to the obsolete word-map of limitlessness, with reluctant concessions to the changed reality.[297]

A *Really* Stark Fact

But events in the early 1970s abruptly (though all too temporarily) jolted our faith in limitlessness. Oil exports to the United States from Arab countries were interrupted (in protest against American policy in the Middle East). In a radio address in February, 1973, President Nixon made a statement to the American people that might suggest he realized the era of carrying capacity surplus was ended and wanted his constituents to realize it too. He said, "We must face up to a stark fact. We are now consuming more energy than we produce." For a nation whose customary word-maps depicted it as "independent" and "the most productive on earth," this was indeed stark, but it was also a monumental understatement of the real situation.

As I explained in an earlier book,[298] a good estimate of the rate at which nature might be replacing the energy deposits man was withdrawing from underground could have been easily calculated. The reasoning was not very elaborate. One merely needed to know: (1) the total weight of the Earth's atmosphere—easily calculated, (2) the fraction of air that was oxygen, (3) how long it had taken for that much oxygen to be released from carbon dioxide (in which it had formerly been bound), and (4) the comparative weight of one atom of carbon to the two atoms of oxygen in each former molecule of atmospheric CO_2.

None of this information was secret or undiscovered; it wasn't even very obscure. Sea-level atmospheric pressure was commonly known, as was the approximate diameter (from which could be calculated the surface area) of the Earth. So the weight of all the air on Earth could be figured to a reasonable approximation with ordinary high school mathematics. Roughly one-fifth of the air was now oxygen, and 99 percent of that free oxygen had been released, it has been estimated, in the last 600 million years.[299] The atomic weights of carbon and oxygen were readily available, and their ratio was simple to compute. So it turned out that about 625,000 tons of carbon had been the *average* amount buried per year in deposits of coal, oil, natural gas, and other less combustible substances since the photosynthetic process began releasing into the atmosphere a net total of one million billion tons of oxygen. Moreover, much of that removal of carbon from the atmosphere had occurred within the Carboniferous period, between 300 and 215 million years ago, so the *present* (or recent) average annual addition to the world's carbon-rich "fossil fuel" deposits could scarcely be even as much as half that long-term average.

By the time President Nixon spoke of a stark fact, the world's human population, with all its technology, was unearthing and burning these substances at a rate that re-oxidized and returned to the air more than four billion tons of carbon each year. Thus the rate of 'harvesting' from this ghost acreage (4×10^9 tons per year) was more than 10,000 times what nature's rate of replenishment could be estimated as now being ($1/2 \times 6.25 \times 10^5$ tons per year). Conservative as the estimate of a 10,000 to 1 ratio might be, that figure had not been calculated in time to deter deep commitment of human societies to such overuse.

The stark fact could be put even more simply. It would have been possible (had it not been for the pre-ecological word-maps) to see how much the output of agriculture and forestry and fishing would have had to increase if *Homo sapiens* were to try to derive more of his current energy expenditures from current energy income (as it is now beginning to be naively imagined we will soon do) rather than from underground resources from antiquity. If mankind is withdrawing annually from Earth's savings about ten times as much energy as we are obtaining from current photosynthetic income (from organic sources), therefore to reduce our dependence on fossil energy *by only one-tenth*, human beings would have to double their use of contemporary photosynthesis. To do that would obviously entail increases in harvested biomass falling somewhere in the almost surely unattainable range between another doubling of yield per acre and another doubling of tilled acreage at existing yields. This unlikely improvement is just taken for granted by those who are today promoting biofuels (such as corn-based ethanol) as the antidote to what even President George W. Bush ultimately called our "addiction" to oil from the Middle East.

To become *completely* free from dependence on prehistoric energy (without reducing population or per capita energy consumption), modern man would require an enormous increase in contemporary carrying capacity—to a total *equivalent* to ten Earths, each of whose surfaces could be forested, tilled, and harvested to the current extent of our existing planet. Without nine new Earths to supplement the one we live on now, it follows that—barring some truly prodigious increase in efficiency—mankind's most "developed" (colossal) ways of life must shrink drastically sometime in the very near future, or else that there must someday be *many fewer* people. Neither alternative, and none of the reasons for them, are contemplated by those who glibly seek "energy independence"—meaning release from dependence on fuels obtained elsewhere than within U.S. boundaries. "Drill! Drill! Drill!" they urge. We are somehow expected to go right on importing energy from *elsewhen*, heedless of any possible termination by nature of the era in which that has been possible.

In the eighteenth century no one could recognize that, by inventing an improved steam engine, James Watt was launching mankind into overshooting the sustainable carrying capacity of this one Earth. Watt has been conventionally regarded as something of a cultural hero for enabling humans to exploit a vast "new" source of energy. He was a clever and decent man who lived in (and exemplified) the Age of Exuberance—the era of optimistic growth of both population and machines. His invention compounded the misconception born of Columbus's discovery of a New World. Burgeoning reliance on fossil energy further augmented the carrying capacity surplus that briefly shaped our ideas, our lives, and our institutions. Watt's invention thus reinforced our naive belief in limitlessness, previously aroused by having gained access to a second hemisphere. He was never taught to think in terms of carrying capacity, and that term from nature's dictionary would remain absent from conventional word-maps until more than two centuries later.

When we consider the fact that a mere one-tenth of mankind's activity the world over was being powered by contemporary organic energy sources, we arrive at a realization of just how stark what President Nixon called a "stark fact" *really* was. Ninety percent of what humans were doing by the 200th anniversary of Watt's invention was being done by withdrawing nature's energy *savings deposits* thousands of times faster than they had accumulated.

A new updated word-map needed to accept the fact that all activity involves use of energy.[300] If only one-tenth of what humans do now is done with this year's income (solar energy captured by contemporary photosynthesis in our crops, timber, etc.), then nine-tenths of what we do *in one year* is done with solar energy that was stored away by processes operating during thousands of prehistoric years.[301]

Triple Peril in Our Time

One of the functions of language, Alfred Korzybski had argued, was to enable humans to become a "time-binding" species—consciously relating present actions to past and future, to heritage and experience as well as to goals. If we are to understand the most pressing dilemmas facing us today, we need to broaden Korzybski's concept of "time binding" so it can take into account modern man's dependence on *ancient* photosynthesis. An updated word-map should make one thing perfectly clear: *Homo colossus* "imports" most of his energy—*from antiquity*. The ratio for "advanced" (more colossal than average) nations is of course higher than the world average. And the fact that Americans import much of their energy from the Middle East is incidental—a problem, given certain cultural animosities, but less of a fundamental threat to our future than the fact of our reliance on *imports from the carboniferous past*.

That reliance has brought us to a time when we face a trio of interconnected monumental dangers to our future. 1. The world has passed the peak rate of extraction of our most advantageous fossil fuels. This is especially evident if we look at the *per capita* annual output. 2. The cumulative effects of the use of fossil fuels thus far have contributed to serious climate change, which is already beginning to modify local environments in many parts of the world. As a consequence of this, 3. The loss of essential habitat has raised rates of extinction of species, including many that humans have used as "resources," or might have found useful in the future. And of course the depletion of various non-biological resources (e.g., metallic ores) is continuing and imperils our future as well.

Public awareness of this triple threat to our future has been delayed by our continued use of familiar but obsolete word-maps. The effort of *Homo colossus* to stay afloat relied on the illusion that a nation could be "self sufficient" by importing energy from antiquity (within its own borders) at many times the rate of current production[302] whereas importing "foreign" fuels must spell disaster. But no nation could ever be "self sufficient" while relying on energy from antiquity. That reliance was profligate, whether we extracted the combustible materials from under our own land or from under foreign lands. By insisting the first war with Iraq was "not about oil" the first President Bush wasted a vital opportunity for enlightening people about our perilous dependence on fossil energy. Had he been a more ecologically informed president he could have used the occasion to redirect his countrymen onto a course toward coming to terms with carrying capacity limits. He could have announced that it was time to begin stretching out our use of such fuels as came from the Persian Gulf area and having them serve *transitionally* toward an inevitably coming era of lower-energy living. He could (and we might say should) have begun the necessary process of enabling *Homo colossus* to revert to being *Homo sapiens*.

"*Modernization*" is part of a word-map that has enabled us to increase our dependence on *past* photosynthesis. Too many technological achievements have boosted a "modern" lifestyle by eliminating any trace of "sustained yield" from our total relation to our resource base. The nations most inclined to imagine that they "have it made," and therefore fondest of the conventional word-map, have been the ones most committed to a way of life that was physically certain to be temporary. Moreover, the commitment of these nations to that obsolete map affects the whole world. Importing from antiquity has given the modern *world* very temporarily the illusion of a carrying capacity several times larger than the world's permanent carrying capacity would be without such imports. So the world's population has continued to grow way beyond the number (and living standard) this planet could realistically support from contemporary vegetation on a long-term basis.[303]

Some Unavailing "Ifs"

Much of what has been said above must seem meaningless or preposterous to readers thoroughly conditioned to traditional word-maps. To those whose predispositions were traditional but who may nevertheless have begun reluctantly to grasp the significance of new insights, it was natural to attempt rebuttal—to seek some way of insisting that the new word-map is at least as faulty as the old, and to reassure oneself that its ominous implications need not be taken seriously.[304]

It might occur to anyone so predisposed that the quantitative argument presented here could be very simply answered in its own terms. If Americans are consuming "fossil fuel" energy thousands of times as fast as it had been stored up by nature's processes, won't it still take thousands of years to run out, since we are drawing from *two-thirds of a billion* years' accumulation? The United States is not yet two and a half centuries old and only recently modernized. Doesn't this mean many more centuries of our national life span are still ahead of us with continuing supplies of fossil energy to fuel them?

The answer to such questions could have been affirmative only (1) *if* all deposits of coal, petroleum, and gas were as accessible to man as the ones already extracted, (2) *if* there were to be no further growth of population or technology, and (3) *if* complete exhaustion of these deposits were the only life-inhibiting consequence of their use as fuel.

Manifestly, none of these "ifs" can be true in the real world.

Regarding "if" number 1: The most accessible deposits have already been used. We are already feeling at least the economic and some of the social consequences of turning to resources that are harder to extract. This issue is confused, though, by political boundaries; instead of facing the reduced geological availability of "our own" deposits, we let ourselves dwell on the human factors impeding availability of oil from Arab countries for our use. When our initial shortage-reducing efforts focused on "suspending import restrictions" and even going to war to "ensure peace and stability in the Middle East," they obscured the fact that the finite stock in the *global* storehouse is being depleted.[305]

Moreover, a substantial fraction of that stock (perhaps the major part) may conceivably be forever inaccessible to us for utterly nonsocial and apolitical reasons. Our old word-maps may have given us faith to remove mountains to get at seams of coal, but the new word-map should remind us it takes prodigious quantities of energy to do so. Much of the world's fossilized energy was stored in forms or places that would require nearly as much, or even more, energy for extraction than would be obtained from subsequent combustion of the extracted fuel. Nuclear physicist Edward Teller (known as "father of the H-bomb") envisioned the use of a thousand underground nuclear blasts a year in Colorado alone to stimulate the flow of natural gas,[306] a resource not easy to extract without some huge effort to fracture the rock strata in which nature had stored it. Traditional word-maps made Teller's suggestion appear to be just another expression of the assumption that "technology will provide a way." From the newer perspective, however, his glib faith can be seen as an indication of the truly exorbitant expenditures of energy that could be required for obtaining energy in decades ahead.

Regarding "if" number 2: "Labor saving devices" are no automatic answer. Many people persist in dismissing the problem of carrying capacity deficits from thought by insisting that technological advancement (which our old word-maps depicted as "inevitable") would surely "keep pace" with our "growing energy needs."

What the old word-maps portray as a solution is recognized by the new nature-based word-map as part of the problem. By *increasing* per capita energy use, technological progress *aggravates* our commitment to living by withdrawing the earth's savings deposits. When the problem-producing rather than problem solving consequences of technology are borne in mind, then it becomes possible to recognize that the world's carrying capacity for people has to vary by its tolerance for their equipment. So with fewer people, the world could accommodate more fuel-using machinery per capita. Or, with fewer items of enhanced technology, the present number of people would have less impact on the world. A future of diminished carrying capacity will compel serious downsizing. The only choice will be: do we reduce the number of people or the number of their resource-hungry tools? Or both?

Regarding "if" number 3: We should remember that an organic system can be fatally ill or mortally wounded even though it is more than 96 percent intact. The removal of mountains of "overburden" standing between us and "the fuels we must have" can seriously interfere with agriculture and other necessary aspects of human life long before absolute exhaustion of the total underground deposit is approached. Also, extraction of any appreciable fraction of the "fossil fuels" stored under ocean bottoms would, as a side-effect, further disrupt the carrying capacity of sea waters for the marine life upon which an increasing portion of human subsistence had been expected to depend (by those who saw in the old word-maps no more than inconvenient and temporary limits to growth). Carrying capacity of the oceans had already been diminished by tanker traffic between the continents.[307]

Also, in regard to "if" number 3, some have long known that carrying capacity on land was already being reduced both by soil erosion and air pollution. "Modernization" increased our power to destroy our lands,[308] and it continues to concentrate *Homo colossus* (with his energy-using activities) as well as some ordinary *Homo sapiens* less lavishly equipped, in cities that constitute a small fraction of the earth's surface.[309] This concentration aggravates in these localities the combustion product accumulations from the fuels already extracted and burned. Such concentration means, moreover, that accumulated heat (and all energy after use ends up as heat) can begin to modify wind patterns, and eventually jet streams and ocean currents,[310] with potentially disastrous impact on existing human communities.

We needed to see long ago that the consequences can become disastrous long before total extraction and combustion of the millions of years of "savings" accumulation could have brought about complete (or anywhere nearly complete) replacement of atmospheric oxygen by CO_2.

For all these reasons, the stubbornly held notions that "There's plenty left" and "We can get it out if prices go high enough to make it profitable" were almost as misleading as the more flagrantly anachronistic use of such words as "unlimited" or "inexhaustible."

Chapter 14

How We Mistook
Destruction for Production

"Our fossil fuels have brought us to a level of abundance and prosperity that was unimaginable a century ago.

"Today they are propelling us forward into a century of disintegration."

—Ross Gelbspan, *Boiling Point*, 2004, p. 61

"Environmentalists expressed 'shock' and 'outrage,' but the energy plan was the inevitable result of placing economic growth on an altar and making all else bow before it."

—Bill McKibben, *The Age of Missing Information*,
1992, p. 109

The semantic malfunction—so prevalent that it accounts for the enormously inadequate recognition of a "stark fact" by a head of state—had been pointed out a generation earlier. William Vogt wrote,[311] "One of the chief causes of our ecologic imbalance is our economic thinking. We identify the symbolic dollar with real wealth . . . . We extract oil, and iron ore, and fine timber, and canvasbacks, and call it *production*."

Using the word "production" in that misleading way as a euphemism for *extraction* was not unusual. Most words have multiple meanings. Context usually sorts them out, and lets us know which meaning applies. This, and the fact that the different meanings of a given word tend to be related, normally enables communication to proceed, but there is a risk of spill-over of one meaning into an inappropriate context. When this happens, the consequences may or may not be serious. In the present instance they have been deadly.

To the farmer, "producing" means "growing a crop"—transforming material substances (soil, water, air) and energy (sunlight) by horticultural methods that manage and channel photosynthesis. In the manufacturing sense "producing" something also means giving form, shape, or being to a product—or "making" that product by assembling components or by transforming raw materials. For the dramatist, to "produce" a play refers to the *presentation* of a work of art to an audience. It involves stage props, actors, scripts, costumes; but the word in this context has less reference to the manipulation of substances except as symbols or representations. Use of the term to refer to *symbolic* manipulation becomes even sharper in a mathematical context, where to "produce" the side of a parallelogram means to project or extend it conceptually. This is the top of the abstraction ladder; no transformation of any substance is implied at all.

When a consumer of manufactured goods, of farm output, artistic performances, or mathematical knowledge "produces" coins from his pocket to pay for a purchase, the meaning is just below the top of the abstraction ladder. The coins are tangible, but the buyers did not make them. "Produce" has become synonymous in this context with *reveal* or *extract*.

It is easy to see how the mathematical and artistic meanings of the word are related to the meaning in a context of farming or manufacturing. But the difference is also apparent so it is unlikely the word will be misunderstood in any of these contexts. On the other hand, it has not been widely appreciated that companies or nations which "produce" crude oil (or natural gas, or coal) or valuable ores do so only in the coin-pulled-from-pocket sense. They extract from the earth a substance that already exists. The substance was *formed* long before—by processes of nature. Substances that were carbon-rich and therefore oxidizable, were rich in releasable energy. The so-called "producer", however, did not put the energy into the substance nor put the substance into the ground.

To use the word "production" to denote extraction has seemed plausible because firms that extract such substances from the earth are as involved with engineering and commerce as any manufacturing concern. But this usage in reference to a process of extraction has enabled us to suppose the process could be expanded as freely as manufacturing and perpetuated as indefinitely as farming. That illusion has been especially mesmerizing to economists, with few exceptions until recently.

Once again, division of labor was involved. Original receptivity to misleading word-maps, and strength of enchantment by them, varied by occupational specialty. Some occupational roles were more legitimated than others by those word-maps. From too many supposing what was untrue, we have all come to grief. All of us, from petroleum prospector to consumer to president, have acted as if the rate at which we could afford to spend our coins was limited only by the rate at which we could extract them from our pockets (or the pockets of someone else). We have expected to be able to use the "fossil fuels" at accelerating rates as long as the discovery of additional deposits also accelerates. This was like a dog's faith in Old Mother Hubbard—supposing whenever and however often she went to the cupboard, she would always be able to fetch from it another bone. But even the rate of discovery of previously undiscovered fossil fuel deposits has peaked and is now in decline, despite much effort by enhanced techniques. By ignoring other constraints we have implicitly assumed that it does not matter by what complex processes Earth's energy wealth was stored away, at what rate the accumulation took place, or how these processes may be articulated with other natural processes that affect us.

Where Is "Away"?

To consider what this means for our future we must take into account both "resource" issues and "disposal" issues, for when we take something from the environment and use it, we transform it, and the resulting substance—even our own bodies when we are through with them—must "go" somewhere. Life in a material world inexorably involves both inputs and outputs, or, as Allan Schnaiberg aptly called them (from the perspective of the environment), "withdrawals" and "additions."[312] We are injecting into the atmosphere, into the waters of our one planet, and onto its lands, greater quantities of end-product and by-product "additions" than the world's ecosystems are capable of reprocessing.

Whatever materials we use, and whatever we make of them, eventually we have something we must dispose of. In this regard, the habits of people in the United States have been derisively summed up by the accusation that Americans have a "throw-away culture." Efforts to encourage respect for the reusability of many things and many substances and to emphasize the need to protect vital ecosystems from contamination were expressed in the slogan "If you're not recycling, you're throwing it all away." The words "it" and "away" were subtly and intentionally vague, to indicate it was future life, not just some commodity, our improvident habits were putting in jeopardy.

For some, the interest in new thoughtways implied by fashionable use of the previously unfamiliar word "ecology" merely led to stigmatizing acts of littering. But if one bothered to think about simple ideas learned in a high school science class about the indestructibility of matter, it should have been obvious that whatever "resources" we extract from Earth in any manner, and whatever products (and by-products) we make of them, we will somehow end up putting somewhere an equivalent amount of "stuff"—in some transformed condition after using up and/or wearing out whatever products we turned those resources into—whether the disposal consists of dumping onto the ground, or injections into the waters of the Earth, or releases into the atmosphere. Foods we eat and beverages we drink, end up becoming sewage, along with various industrial effluents. Fuels we burn end up being "greenhouse gases," as we have at last begun to recognize.

None of these facts mattered so much when this planet had only a few million, or even a few hundred million, widely scattered human inhabitants. They matter greatly now that we number more than six billion, and now that so many have been technologically-culturally transformed into *Homo colossus*. In a less densely populated world inhabited by *Homo sapiens*, there was plenty of "away" for our discards. In an overloaded world of *Homo colossus*, throwing things away, even our own bodies when we are through with them, becomes a real problem. Even if population were stable, with a balance of birth rates and death rates, burials would be cumulative generation after generation. Already there is competition for land between cemeteries and other land-uses such as airports and highways.

So in addition to having to outgrow traditional word-maps enough to become concerned with finite resources as a carrying capacity limit, we have come to a time when the disposal side of life's processes on a finite planet with finite oceans and finite atmosphere has to be reckoned with. The things we have been doing to those parts of our home planet have been accumulating to the point where they now affect our lives. Changes human actions could formerly inflict with impunity upon the biosphere when we were few and "underdeveloped" become sins against posterity when we are "developed" and number ourselves in billions.

Clinging to Weak Foundations

It must be said again: the quest for better word-maps arises from recognition that old language habits inaccurately depict reality. There *is* a real world with which we have to keep trying to come to terms. But it must also be said again: a new vocabulary and enlightened thoughtways, more accurately describing that real world, could not guarantee that the existence of six-plus billion human beings upon it can be utopian and perpetual, with never a need for our numbers or our affluence to stop growing or begin to decline. The imaginary world capable of magically supporting ever-increasing populations with ever rising per capita resource appetites and environmental impacts is not the real world. The real world is finite. It has limited carrying capacities. The trajectory of modern civilization has been less simple to calculate and extrapolate than the path of a batted baseball, but learning to foresee where it is headed is vastly more consequential.

Acceptance of a realistic word-map implies disillusionment—dis-*illusion*-ment. It is essential at long last to see and acknowledge some important implications for our beliefs about each person's relations to his or her fellow humans that do necessarily follow from abandonment of illusions about the nature of humankind's relation to an environment recognized at last as finite and having palpable biogeochemical features. It will become apparent as we grow accustomed to the new worldview that full attainment of love-thy-neighbor human brotherhood is impeded not only by psychological or cultural obstacles but also by obstacles of a more geophysical nature.

The countries we traditionally called "backward"—those living mostly on *present* solar energy—learned to desire very earnestly to "catch up" with those whose modern technology had given them an ability to devour the past. This is why, at the first United Nations Conference on the Human Environment, held in Stockholm back in 1972, the already belated efforts of industrialized nations to begin protecting this heavily populated planet from the consequences of industrialism met resistance from the yet-to-be-industrialized countries. Their plea that "Environmentalism must not stand in the way of development" had a plausible sound to anyone still following obsolete word-maps. But the ability of delegates to that conference to be diplomatic and achieve some verbal reconciliation of their opposing perspectives did not make it any more realistic than before to imagine that a European or American standard of affluence for the entire world was an attainable goal. Likewise, the Kyoto conference in 1997, which produced a seriously less-than-adequate international agreement to begin reducing greenhouse gas emissions to forestall problems that will arise from global climate change, had to make crippling concessions to countries still aspiring to become modern. The obsolete word-maps supported attitudes that were just too powerful to supersede.

Humanitarian attitudes to which many of us have tried to adhere have caused us to imagine sometimes that conflicts of interest between nations have no real basis in nature, that they arise only from ethnocentrism or hyper-nationalism, or from the historic vestiges of previous (and equally unnecessary) conflicts. Noble sentiments such as these have made it hard for us to face a fact that should now be apparent—there is a *real* conflict of interest between nations devouring resources thousands of times faster than the rate of natural accumulation of those substances. They compete for shares of a limited pie. Their efforts to grab ever bigger slices of the *vanishing* pie must be especially galling to people in countries not yet privileged to have arrived in the upper region of the resource-consumption range but who have already learned to covet the rewards of such "good fortune." Struggles to keep on taking the earth's resources, to acquire the ability to take them, and to keep them from being taken, will doubtless intensify human conflict in the twenty-first century. Those obsolete word-maps have blinded us to the reasons for this. Continued reliance on past illusions will at least *enable* a bad situation to worsen. Refusal to be disillusioned likely *ensures* disaster.

The weakness of the foundations of optimism, as exposed by a new perspective, became apparent many years ago to those with informed vision. In 1908, for example, addressing a conference on conservation, Theodore Roosevelt[313] praised the growth his country had attained by its "lavish use of our natural resources" but went on to say that "the time has come to inquire seriously what will happen when our forests are gone, when the coal, the iron, the oil, and the gas are exhausted, when the soil has been further impoverished and washed into the streams, polluting the rivers, denuding the fields, and obstructing navigation." Now, a century later, there are many, still navigating by obsolete word-maps, who cannot seem to comprehend that Rooseveltian concern.

As long ago as 1929, Robert and Helen Lynd sounded like the 1970s when they wrote, in their study of a newly industrialized community in Indiana which they called Middletown,[314] that when timber still stood on the banks of its White River in 1890 it "was a pleasant stream for picnics, fishing, and boating," but it had since shrunk "to a creek discolored by industrial chemicals and malodorous with the city's sewage. The local chapter of the Izaak Walton League aspires to 'make White River white'" they said.

April 22 each year has become "Earth Day" to mark the anniversary of the 1970 "birth of the modern environmental movement." In September 1969, at a conference in Seattle, U.S. Senator Gaylord Nelson of Wisconsin first proposed a nationwide environmental protest effort to get "the environment" onto the national agenda. The following spring some 20 million Americans participated, advocating achievement of a healthy, sustainable environment. The national coordinator Denis Hayes and a young staff organized massive rallies all across the nation. Organized protests against environmental deterioration occurred on many college and university campuses. Groups fighting oil spills, factory and power plant pollution, toxic dumps, pesticides, wilderness desecration, and wildlife loss coalesced around the common issue of sustainability. Twenty years later, Earth Day 1990 helped prepare for the 1992 United Nations Earth Summit in Rio de Janeiro by putting environmental preservation on a global agenda.

A generation ahead of his public, after the U.S. had *exported* substantial quantities of petroleum products to sustain various allies in the Second World War, and before Americans became so dependent upon crude oil from elsewhere on this planet, William Vogt wrote:

Our most prodigal wastage is, perhaps, of gasoline. We are an importing nation; and every day we waste hundreds of thousands of gallons. All manner of drivers let their motors run when they are not in use. Our tensions find outlets in racing motors and in traveling at high speeds that reduce the efficiency of our cars. We build into our automobiles more power and greater gas consumption than we need. We use the press and radio to push the sales of more cars. We drive them hundreds of millions of miles a year in pursuit of futility. With the exhaustion of our own oil wells in sight, we send our Navy into the Mediterranean, show our teeth to the U.S.S.R., insist on access to Asiatic oil—and continue to throw it away at home.

Despite all such warnings, and in the face of history, some capable writers continued even into the 21st century vigorously denouncing the unwelcome view that a finite earth has real limits which we have ignored at our imminent peril. One publication that attained international notoriety for such a rosy view was by a Danish statistician, Bjørn Lomborg. His ostensibly "realistic" challenge of mainstream environmentalists' concerns had developed after he had embraced the almost pathological optimism expressed by an American economist, the late Julian Simon.[315] An epigraph placed in the front of Lomborg's book consisted of the following patently unecological and demographically naive word-map by Simon:

This is my long-run forecast in brief:

The material conditions of life will continue to get better for most people, in most countries, most of the time, indefinitely. Within a century or two, all nations and most of humanity will be at or above today's Western living standards.

I also speculate, however, that many people will continue to *think and say* that the conditions of life are getting *worse*.

Not surprisingly, soon after its publication a number of eminent scientists wrote informed articles rebutting Lomborg's book.[316]

An earlier defender of the traditional "growthist" faith was John Maddox who edited *Nature* for 22 years but stoutly adhered in the eighth decade of the 20th century to the conventional word-maps, including their confusion of money with wealth. Maddox wrote:[317]

> The usual distinction between renewable and nonrenewable natural resources is unfortunate because it is clear by now that the proper exploitation of natural resources is governed much more by economics than by the simple arithmetic of how much food can be grown with how much sunlight, or how great (or how small) may be the amounts of particular minerals locked up in the earth's crust.

Is it really *a good thing* that our exploitation of natural resources is today "governed by economics" and not by serious consideration of the quantities that exist of the substances we use and depend on continuing to use? Can we really afford to suppose our species has become altogether exempt from constraints of nature?

As the Lynds had pointed out early in writing about Middletown,[318] mankind typically makes "as little adjustment as possible in customary ways in the face of new conditions . . ." Maddox epitomized this in his insistence that the present time "appears to be one at which forecasts of scarcity are less valid than ever."[319] He even said, "Famine is not a threat but a scarecrow . . ." and ". . . many of the hungry nations of the world are on the threshold of unaccustomed plenty." He viewed as an illusion the idea that exhaustion of certain materials could spell civilization's collapse. "Minerals are now more plentiful than ever, whatever the more distant prospects," he asserted. "The threat of a scarcity of energy, real enough in the 1950s, has already been dispelled," said Maddox in 1972 (virtually on the eve of the Arab oil embargo).

Homo "sapiens"?

Thus have members of an intelligent species, one that has learned to import energy from the past and thereby proliferate far beyond the present or future carrying capacity of its planetary habitat, tried to reassure one another. They insist that the billions of human beings now living (with still more to come) can realistically aspire to live as lavishly as the most prodigal few hundred million have recently been doing. This they suppose is their future, on a planet that is finite and whose resources and humanly useful features were provided by processes of nature now being undone by these humans at rates that are thousands of times faster than the rates at which those habitat-creating processes originally occurred.

It is important to remember how we became human. *Homo sapiens* acquired a large brain because we had hands and vocal apparatus whose potential advantages could become actual only by further development of neural control mechanisms. Our not-yet-quite-human ancestors, those early hominids, had useful hands not because their descendants might someday want to operate computer or piano keyboards or manipulate surgical instruments or catch and throw batted baseballs, but because *their* ancestors had been tree-dwelling brachiators. So the brains we have are likewise products of our past, not of our future. Sadly, but truly, their powers of foresight may be inadequate to forestall, or cope with, the future (the "fate") they have wrought. Even our brains' contents (whatever we happen to have learned) continue to be shaped more by past experiences than future prospects. Hazardous as that may be to our descendants, it is a fact.

Three decades ago, in *Overshoot*, I tried to show readers how mankind became locked into stealing ravenously from the future. We had unwittingly become indirect antagonists of our own posterity. Oblivious to carrying capacity limits, we exhausted the time in which it hardly mattered that overshooting carrying capacity seemed equivalent to enlarging it. Today that prodigal misperception has become a grievous error.

When the world fell into economic recession late in the twenty-first century's opening decade, with losses of millions of jobs, foreclosures of home mortgages, major investment company collapses and bank failures, everyone believed the most urgent task for people in power (or would-be leaders), was to restore high rates of business activity. The cry and the promise: "Get the economy back on track!"

But the "track" we had been on was ravenous overuse of a finite planet. Unlike earlier recessions *prior* to the exuberant growth of population and industry in the modern era, *this* time the abrupt downturn from economic prosperity occurred in a world whose once-enormous stocks of nonrenewable resources are now seriously depleted. It is a world already recognizably damaged by disposal of the products of lavish exo-metabolism by *Homo colossus*. This time economic "recovery," if pursued in traditional ways, is either bound to be impeded by the unrecognized carrying capacity deficit, or will so further worsen that deficit as to hasten its ultimate catastrophic consequences.

Ultimately, the carrying capacity deficit must trump financial recovery. We have made "ultimately" a lot closer than most people recognize. Posterity will not thank us for frantically returning to our customary stealing ravenously from their future.

CHAPTER 15

How Our Future Is Bottlenecked

Humanity has managed, primarily through the invention of agriculture, to increase the carrying capacity of the planet for people. Further, by exploitation (and destruction) of a one-time bonanza of resources—such as fossil fuels, concentrated mineral deposits, deep agricultural soils, extensive underground stocks of fresh water accumulated over hundreds of thousands of years, and diverse other organisms—humanity has increased its population size beyond a level that is permanently sustainable. This inevitably means that death rates will rise . . . .

—Paul R. Ehrlich, *The Machinery of Nature*

"Look at how we live . . . . We live the way we live, because you live the way you live."

—Alejo Suarez, Ecuadorian inhabiting a horribly degraded environment. Quoted in Bergman & Fontana, OpEd: Oil's Human Cost

Our species is substantially confined to an ineluctably finite global habitat, where resources will not suffice to assuage universalized and perpetually escalating desires. So a revolution of rising expectations should have been expected to nurture a revolution of rising despair. A growing sense of "oppression" has had *ecological* roots—less visible, perhaps, but more inexorable than the ideologically touted tyrannies of ruling classes or regimenting activities of overzealous bureaucracies. As people have reluctantly begun to sense that the "impossible dream" of universal modernization probably *is* impossible, this may have helped foster an epidemic tendency to resort to violence—as a means of denying unwelcome truth.

In the twentieth century, human numbers exploded as we occupied vast new niches made available by advances in technology and organization. It was a period when a very substantial fraction of the world's total *Homo sapiens* population committed itself to living as *Homo colossus*, and millions of other people in many lands aspired to follow in their seven-league boots footsteps. We swarmed into the ephemeral niches made by and for *Homo colossus*. Almost no thought was given to the possibility that those gigantic footsteps portended an awesome and deepening carrying capacity deficit, with the new niches being irreparably temporary—based as they were on ravenous use of nonrenewable resources and upon virtually unfettered spatial expansion of human activities.

Biologists could describe many past instances where one species or another has expanded into a newly available niche, overshot the permanent carrying capacity of that niche, found itself going through a "bottleneck" of drastically narrowed life opportunities, and therefore experienced a die-back to a number that was commensurate with the diminished carrying capacity residually available to sustain it.[320]

Because humanity's enormous technological accomplishments in the twentieth century turned so many of us into resource-ravenous *Homo colossus*, the twenty-first century will have to be a bottleneck era for the world's human population.

Earth cannot continue supplying the resources *Homo colossus* has learned to consume so voraciously. Nor can it absorb (and recycle) the prodigious quantities of noxious, toxic, landscape-altering or climate-changing end-products injected into the global environment by our somatic and especially our exosomatic metabolism. Accordingly, the number of humans living upon this planet, although still increasing as I write this, will very probably be markedly fewer by 2100 than the global population of just over six billion at the start of 2001. A further challenge will be the need for people accustomed to economic growth (myopically called "progress") to adapt to inevitably squeezed and constrained "standards of living" in a resource-depleted world.

But as the century began, too few people seemed yet to realize there was a bottleneck in their future, nor was there enough awareness of how horrendous will be the passage through it. The vaunted "land of opportunity" was becoming a land of exhausted and constricted opportunities. Wars were going to be fought among competitors over access to dwindling resources. Scapegoats would tend to be be sought—to serve as an "explanation" of the self-inflicted miseries resulting from human overuse of the planet. Comparatively innocent scapegoats would become targets of malicious and destructive actions that will occur and recur.[321] Wars of Mass *Distraction* ("WMD") have already commenced. The eventual subsidence of hostilities will lag appreciably behind exhaustion of ill-conceived provocations. Various ideological rationalizations will be ventured to "explain" and "justify" the vicious onslaughts by one population against another among our conspecifics.

Happy Pre-Bottleneck Moments

Portions of several chapters of this book were written on a so-called laptop computer which I took with me when my wife and I journeyed to New Zealand to become acquainted with our first great-grandson, born in the previous year to our Kiwi granddaughter and her Australian husband. This portable computer, together with the Boeing airplanes in which we flew, and the structures and organizational features at the several airports we passed through, all served once more as reminders of our dependence on today's many specialized occupations that lurked unseen (but appreciated).

It was a trip that incidentally afforded us additional personal evidence that we live on an approximately spherical planet, as we took off on the first official day of spring and landed on the first day of autumn—after being aloft for about 12 hours, not for six months. Traveling via commercial jet from the northern to the southern hemisphere made advancing two seasons overnight quite easy. The comfortable trip from the North American west coast to the South Island of New Zealand showed us most vividly once again what seemingly magical achievements are made possible by extreme division of labor.

The visit was an absolute delight—though we saw less of New Zealand's scenery than on previous visits. At age nine months our great-grandson's most obvious conditioned reflex was a big smile automatically turned on whenever he heard the gentle whir made by my digital camera's lens extending to its picture taking position. We were on the ground in New Zealand for only three and a half weeks, during which this new little member of our species (and of our extended family) entered the quadrupedal stage of development. He quickly became enthusiastically mobile, learned to let himself down feet first off a bed or sofa, cautiously probing with one foot to feel the floor's location, before completing the descent. Once down on hands and knees he would quickly crawl to where he could grasp a chair or table leg and pull himself up to a standing position (in implicit anticipation of becoming bipedal months or weeks hence).

He was fascinated with buttons (and with observing the hands of his parents, grandparents, and great-grandparents doing things at a computer keyboard). These interests led during our visit to his differentiating his own right index finger from the rest of his little digits. Whenever his eyes happened to focus on a button, that infant index finger reached out to push it—usually to no effect (with buttons on clothing)—but when it happened to be the button on my talking digital watch he seemed entertained by the effect of pushing it, which caused a little electronic voice in the watch to speak the hour and minute.

Brief though our visit was, it was a sufficient span of time to observe his happy vocalizations change from merely exuberant and variously pitched "aaaaahh" or "uuuuuhh" (primordial singing?) to real babbling—imitations of several frequently heard consonants—on the verge of making attempts at voicing potentially meaningful syllables.

Many digital photos of his infectious smile were downloaded from the camera into the laptop computer to be cherished after our return home. Day after day, and even year after year, the brilliant on-screen images will inspire hope that by the time surviving members of his generation have emerged from the coming bottleneck, when *he* may himself have somewhere a great-grandson he will wish to visit, somehow his contemporaries will have attained the wisdom Linneus implied was characteristic of our species when he named us *Homo sapiens*.

Omens of the Post-Bottleneck Future

People who travel abroad often purchase picture postcards. We did so on this trip. One was a spectacular photo of New Zealand's highest peak, called "Aoraki" by the Maori people, meaning "the cloud piercer." It was renamed more prosaically "Mt. Cook" by European settlers of New Zealand after the English explorer, Capt. James Cook. In the postcard photo it even more emphatically manifests today the Maori conception than it formerly did, as the sharp asymmetrically pointed summit juts up into the sky like an inverted feline fang. Its height above sea level is now given as 3,754 meters, or 12,316 feet. The reason it looks so sky-puncturing is that some years ago a huge rock slab that *was* the summit, upon which intrepid mountaineers used to set foot, split off and crashed down the mountain's flank, leaving the present topmost snow cornice a good ten meters lower. Future climbers, however earnest their zeal and however rugged their muscles, simply cannot ever again stand where climbers once did. So (like Mt. St. Helens, in Washington state, the volcanic cone which my little great-grandson's great-uncle summited as a teenager—before it lost *its* former top in the 1980 eruption), Mt. Cook stands as another stark reminder that goals once attainable can become utterly impossible experiences to replicate.

What are the chances that *Homo colossus* will undertake the necessary downsizing in time for there to be a decent future for *Homo sapiens*? And what are the chances that *Homo sapiens* will exercise sufficient reproductive restraint that our numbers will voluntarily decline sufficiently (and soon enough) to eliminate the carrying capacity deficit we ran up in the twentieth century? Will the bottleneck's full miseries have to happen, or will people learn enough self-restraint to ameliorate them appreciably? To ask these questions is almost to imply the saddest of answers.

Before we were launched into industrial-level division of labor, the geographic history of *Homo sapiens* already had divided us into several visibly distinguishable races, and many different cultural groups. People adhere to an assortment of diverse religions (each tending to regard itself as uniquely true, and often assuming itself entitled—if not obliged—to indulge in conversion—or sometimes even violent eradication—of rivals). These prior divisions do not bode well for the degree of global cooperation we now *need* to achieve.

In September, 2001 in southern France, a tour bus filled with American tourists, including my wife and me, en route to Nice, suffered a breakdown and pulled to a stop at the side of the motorway, where the driver phoned his company for a replacement bus to come and pick up his passengers so we could continue our journey. Meanwhile, he got out a tool box, opened up the engine compartment and made vain efforts to correct the problem with his bus's transmission. Some tourists waited patiently in their seats. The interior of the bus soon became overly warm so many of the passengers chose to climb out and stand in the open air beside the road. There they chatted casually. Presently along came a French police car and a pair of officers got out of it to see what was our problem. After learning of the mechanical difficulty from our driver and the fact that a replacement bus had been called for, and acknowledging that our bus had been pulled far enough out of the traffic lane to comply with legal requirements, they proceeded to try, with virtually no English, and with our group's even more negligible French, to communicate to us the jolting news from New York and Washington on that day—forever after known simply as "9/11."

What developed then was an interesting lesson in the fact of quasi-speciation. We English-speaking foreigners and the two French officers were all members of the single species, *Homo sapiens*, but the language barrier between us was considerable, reducing communication toward the level of information exchange between two separate species. From gestures and a few cognate words common to both languages, we did manage to piece together that these officers were telling us something about (1) an airplane crashing into a tall building in New York, (2) something significant having happened at the Pentagon, near the U.S. capital city, and (3) that somewhere a plane had been hijacked. We did not know whether these three news items were somehow connected or were just an ordinary day's panoply of unrelated mishaps. What filtered through the boundary between two languages was a rather garbled and incomplete story of that day's induction of the world into an era in which there would soon be steady insistence that "everything has changed."

After an hour or so a replacement bus arrived, and our group's luggage was transferred to it. We tourists climbed aboard, continuing our conversations the rest of the way to our hotel in Nice. These exchanges of thoughts now turned to the kind of speculations that typify efforts by members of the speaking species to put words together to serve as plausible gap-fillers in our fragmentary knowledge about something we sense to be momentous. The speculations voiced by this busload of tourists led to consensus that perhaps the hijacked plane and the one that had crashed into the tall building were one and the same, but until the bus arrived at the hotel there would be no access to further information to check this inference, or any others. The group speculated further that the hijacker must have somehow come to be at the plane's controls. Several individuals expressed confidence that any legitimate airline pilot would have been capable of avoiding a tall building when flying on a clear morning over New York. But the idea that a hijacker at the controls had *deliberately* hit the building, or that intending to do so had been the reason for the hijacking, did not yet occur to these speculating tourists.

A bigger gap in knowledge about the incident remained unfilled. We knew somehow the Pentagon figured into the day's news, but that piece of the nebulous puzzle was too isolated to mean much yet. However, it did prompt some opinionated discussion of airline "security procedures." Speculative comments quickly led to substantial agreement among this busload of tourists that the passenger "screening" procedures at U.S. airports at that time were worse than inadequate—they were "pathetic." But the conversation scarcely touched on ways of making any genuine improvements. Nobody's comments aboard the bus provided any realistic preview of the inconveniences that would change air travel in the days and years ahead.

When we arrived at our hotel for that night, everyone was more interested in turning on the televisions in their rooms than in getting ready for dinner. Newscasts in our own language on CNN brought closure to some of the information gaps so inadequately filled by our speculative verbalizations aboard the bus. Even as the immediate picture of the 9/11 events grew clearer, though, few of us at dinner that evening expressed any realization that those events would lead to a war that would cost more American lives than were lost in the collapsing World Trade Center towers and the burning portion of the Pentagon (and to inflicting vastly greater non-American mortality in coming years). Such words as "We can't let them get away with it" seemed the nearest thing to a conclusion drawn that evening by these dining American tourists.

Not Altogether New

This was hardly the first encounter of commercial aviation with the actions of people so disaffected that they become "terrorists." In March, 1977 in the Canary Islands, a KLM 747 and a Pan American 747 landed at Tenerife. The two jumbo jets had been diverted from another airport where terrorists happened then to be interfering with flight operations. At Tenerife, fog was closing in, and as the two diverted and refueled planes prepared to resume their transoceanic flights, they collided on the fog-obscured runway. Five hundred and eighty-one innocent travelers met death in the fiery mishap.

That was a record toll at the time for a commercial aviation disaster. A similarly civilization-challenging shock *could* have come four years earlier than that. In 1973 a comparable number of airline passengers might have been blasted out of the Italian sky to publicize grievances not necessarily shared by any of them but passionately felt by one of the groups of frustrated people so ubiquitous in the modern world. It did not happen, though, because portable heat-seeking anti-aircraft rockets in possession of Palestinian terrorists had been confiscated by authorities at the Rome airport in time to prevent such an incident.[322] Somewhere else, at some other time, such a disaster would occur—because *Homo colossus* had made this a world in which too many members of the human species who now suffered *significance deprivation* had both the tools for wreaking havoc and an overwhelming urge to declare to the world, "We exist and we matter."

Temptations of 21st Century Technology

Modern terrorism is both shaped and invited by modern circumstances. Airline hijacking, which would become a tactic of international terrorism, was obviously unavailable as a tool of militancy until commercial aviation came into being. Once airplanes had been invented, however, it was natural that they should undergo steady improvement on several dimensions, with their range, speed, and size of payload all being impressively increased. The engineers who accomplished these improvements—symbol-wielding humans all—could not be expected to foresee what opportunities and inducements they were providing to sufficiently desperate individuals and groups who might be tempted to practice a new form of piracy. Hijacking did not become common until airliners became huge, swift, and long-range. Only after all that impressive technological and organizational progress did it become possible for a commandeered airliner to be diverted to virtually any destination the terrorist might choose.

No matter how "inhuman" the actions of terrorists may seem, such people *are* people. The human organism being what it is, not until the jet age could an airplane be expected to reach some remote part of the world soon enough for the hijacker's desperation to remain focused all the way upon political or quasi-political goals, without yielding to preoccupation with his own bodily needs for sustenance or sleep. Not until aircraft grew large enough to carry passengers by the hundreds rather than by the dozen, in a world where life for some had lost its luster, was there substantial probability that someone on board would harbor some motive stronger than his own instinct of self-preservation. And then, of course, the impressive number of hostages he could hold in jeopardy by threatening destruction of the one aircraft provided substantial leverage for an act of political blackmail.

The technological advancements represented by jet age commercial aviation were produced by decent, law-abiding members of the symbol manipulating species, but these developments have made potentially available for use by embittered members of that same species an unprecedented arsenal of destructive devices and tactics for asserting their existence and significance. Moreover, modern communications media, also created by good people, have enabled the terrorist to command by his misdeeds the attention of far-flung millions—and perhaps to work his nefarious will upon whole nations instead of just upon the hostages in his immediate presence whom he threatens directly with physical violence.

Modern media of communications are thus subject to being used in ways never contemplated by (and perhaps inconceivable to) the authors of the first amendment to the U.S. Constitution who sought to guarantee a free press. New circumstances thus raise unavoidable questions. Is it possible, for example, to prevent unscrupulous perpetrators of violence from taking illicit advantage of the existence and nature of the mass media of communication? If it is possible to deny such groups access to the media, can this be done within the spirit of the first amendment? The question of constitutional acceptability is for legal minds to consider, but the question about feasibility of prevention is essentially technical, to be answered in the best available light of social science knowledge. To bring that type of knowledge to bear we must consider some things that are known about human beings in general, and about social movements and terrorists in particular. Then, taking into account some trends in terrorism, the media's role in light of these factors can be examined.

Terrorism and Its Antidote: A Superficial View

Terrorism has raised serious legal and ethical issues, and questions of policy. Resolution of these issues, and answers to these questions of policy may depend on what factors are assumed to constitute an explanation for the terrorist episodes that get our attention. Two members of the editorial staff of *Reader's Digest*, who wrote an article for that magazine a quarter century before the 9/11 episode, provided an example of this connection between assumptions and answers.[323] They purported to discern three terrorist types: (1) "rootless rebels" who "invariably" believe devoutly in "a fairy-tale ideological world of good guys versus bad guys;" (2) rootless members of ethnic minorities "goaded by an outraged sense of injustice;" and (3) common criminals. Their article ended by recommending congressional authorization of FBI wiretapping for anti-terrorist "intelligence collection." They quoted Aleksandr Solzhenitsyn as an advocate of firmness in suppressing hijacking and other forms of terrorism. According to that pair of *Reader's Digest* authors,

> Ultimately . . . the only truly effective counter-weapon is *intelligence*. That means—in the United States—giving the FBI the legal and scientific tools it needs, plus public understanding and support for their aggressive use. It means spies, networks of paid informers, wiretaps, bugs, computerized dossier systems—the whole spectrum of clandestine warfare so necessary to the cause, yet so vulnerable to attack by civil-libertarian extremists.

It is particularly sad for me as a retired educator to realize how commonly we now use that word *intelligence* to mean merely accumulated information (often unreliable) obtained by one form or another of spying upon an enemy or potential enemy. The dictionary I most often consult gives as its first two definitions of the noun, intelligence: "1. the capacity for understanding; ability to perceive and comprehend meaning. 2. good mental capacity . . ." Those definitions make it a word applicable to the nascent traits in my great-grandson that qualify him as a newly launched member of the species *Homo sapiens*. They denote a quality in my students that made university teaching a joy rather than merely a source of income. It was *that* characteristic of our species that has been the most marvelous legacy of all those selection pressures that cumulatively differentiated us from our chimpanzee and bonobo cousins. Only subsequently does my dictionary define the word as designating "military information about enemies." Sadly, that seems to be the more familiar usage today.

Terrorist Behavior as a Social Phenomenon

To put terrorism in better perspective, it is essential to consider some basic insights into the nature of social movements. Even those social movements that resort to terrorist tactics are products of the social nature of *Homo sapiens*. So if we want to understand terrorism we must draw upon basic knowledge of the sources and characteristics of human sociality.

Human beings have to be social to survive. As a species, we are helpless in infancy and we appear to have few inborn patterns of behavior that could enable us to cope as lone adults. So we develop the very traits that make us human as a result of the myriad interactions we have with others, especially as we are growing up. Above all we become communicators; we learn language. Communication looms larger in human life than in the life of any other species, even the most social of the insects, birds, or pre-human mammals.

In these basic respects, terrorists are human—"inhuman" as their actions may seem to those of us who regard ourselves as law-abiding. At least some terrorist acts need, therefore, to be viewed as (lamentable, desperate, incoherent) attempts to communicate. To some extent terrorists are people for whom more conventional means of communication seem unavailable or ineffective. In some instances, at least, the conventional means *seem* unavailable because they *are* unavailable. Sometimes, however, appearances are deceptive.

Having acquired human traits and skills from participation in groups, continued group involvement becomes for all of us a major drive. A normal human response to the group incubation of human personality is some sort of loyalty to the group(s) with which we identify. More or less unquestioning preference for in-group thought ways and behavior patterns, and aversion to out-group thought ways and behavior patterns (when they differ from familiar ways), are also normal. In a world of frequent inter-group encounters, of course, serious frictions between groups can result from such altogether natural in-group bias. Genetically, we humans are so constituted that we are destined to spend our lives absorbing from our associates a non-genetic heritage (i.e., culture). As that kind of creature, we are naturally and almost inescapably ethnocentric. Some instances of terrorist ideology and behavior need to be understood as (perverse) manifestations of such humanly normal group loyalty and ethnocentrism.

For most of us, these drives (to communicate, to identify with our group, to reject alien ways) arise and express themselves first in the context of that kind of group we call a family; but as we mature, we acquire interests that can be implemented by other sorts of groups. Our ethnocentrism may become somewhat tempered with fascination for the exotic. One result of human maturation is the formation of voluntary associations, or clusters of people who share interests somewhat more segmentally than the way interests remain all encompassingly mutual in the intimacy of a family.

Now a distinction needs to be made between two broad categories of activity. Some voluntary associations engage in activity that is mainly consummatory—concerned with members' interest in self-gratification. The behavior is indulged in for its own sake, rather than as a means to some more ultimate goal. (A book-discussion group, or a camera club, or a clique of gourmet diners or wine connoisseurs, would be an obvious example of a voluntary association that was mainly consummatory.) In other associations, the principal activities are more instrumental; the behavior of the members is addressed to the pursuit of some purpose other than direct personal pleasure. Gratification may be a by-product of such activity but it is not the aim. Labor unions, political parties, and professional associations, such as the American Bar Association (ABA) or the British Association for the Advancement of Science (BAAS), tend to be of this instrumental type.

For some members even in groups like the ABA or the BAAS, organizational activity as such may be especially gratifying. The line between instrumental and consummatory activity can thus become blurred. Recognition of this fact can help us to understand variations in motivation toward terrorism. For some individuals in some circumstances, *participation* may become a goal in itself.

Sometimes an association that originated from consummatory interests takes on a more instrumental cast (as when a fraternal organization provides insurance and other services for its members, or when it becomes a pressure group seeking to influence public policies). It is important to recognize, however, that the drift can be in the other direction, from instrumental to consummatory. We have to keep this in mind to avoid making misleading inferences about what makes terrorists tick. Their violence may not always be as goal-directed as it purports to be.

A voluntary association whose goal is to remedy a condition perceived as a social problem, by bringing about change in a larger society, is called a social movement.[324] Some social movements only seek to change public opinion. Others actually strive to change behavior, or to restructure social relationships. Insofar as the group's actions are calculated to serve these ends, the movement is instrumentally oriented, but actions of an organization that are ostensibly instrumental may in fact sometimes be more nearly consummatory. For example, when some group trying to promote a lost cause vainly continues its activities, the interest it really serves is its members' need for reassurance of their own enduring virtue and significance. Their ostensibly instrumental behavior is actually "expressive." It expresses their earnest desire for self-respect. In many instances it may be that kind of desire that underlies acts or threats of violence by terrorist movements.

James Q. Wilson suggested that, for many of the people involved in the race riots of urban America in the 1960s, those riots were *expressive* acts, satisfying to the extent that they gave expression to a state of mind.[325] As he further suggested, whereas the previous generation had been absorbed by notions of how in existential terms one might justify an effort to assassinate a tyrant, later youth had been impressed with the argument that violence, when practiced by the wretched and oppressed, may be *intrinsically* valuable—*as an assertion of self.*

What Wilson said of the racial violence in an earlier decade probably applies more generally to contemporary terrorism. Terrorist action may be rewarding to the militants who engage in it because it serves as an assertion of self in an age that oppresses even middle class youth with what might well be called "significance deprivation." In a city of several million inhabitants, or anywhere in the non-dominant nations on a planet overwhelmed by the machinations of *Homo colossus*, almost anyone can wonder at times, "Do I, as one individual, really matter?"

Writing about the sometimes violent activism in the 1960s by university students who were so largely of white middle class origin, John W. Aldridge emphasized the expressive element:[326]

> They were born twenty years too late to have a part in that knightly crusade against tyranny which World War II now seems sentimentally to symbolize for their fathers . . . .

Activism, perhaps even terrorism, provides a substitute for that lost opportunity. It enables other people in another time to fight their own "morally acceptable war," their own knightly "crusade" (or for Islamists, counter-crusade) against tyranny—as each group sees it.

Opportunities and Incentives

"Why now?" What is it about the state of our world today that enables (and provokes) human beings to terrorize other human beings? Answers to that question must involve consideration of both the changing opportunities for, and changing compulsions toward, terrorist activities. Unless the changing structure of opportunities and compulsions is taken into account, common sense or Readers Digest assumptions about the causes and motivation of terrorism are likely to give rise to quite ineffective remedies. The so-called "war on terror" may need to be written off as expressive more than instrumental. Opponents of terrorists may embrace their own fairy tale, one that prescribes no better strategy than warfare by the good guys, sometimes clandestine (i.e., the spies, paid informers, wire-tappers) or, when the occasion allows, overt (send army brigades, marines, national guardsmen and reservists, and disseminate slogans—e.g., "We support our troops").

Clearly, terrorism does not arise merely in response to economic hardship or political repression personally experienced by the individuals who participate in these movements. In view of the middle or even upper class origins of many militant leaders (and even of the astonishingly numerous suicide bombers since 9/11), it is not material deprivation that motivates them. Especially insofar as terrorism may be expressive, it arises from subtler forms of deprivation—deprivations associated with being born too late in the world's history.

Much of the culture that most of us have internalized was formed in a world very different from the world of today. The change in circumstances between then and now bears heavily upon the social movements of our time. A darkening future may foster movements that are less instrumental and more expressive than their members suppose. This may be a crucial consideration for understanding the roots of at least some of the publicity-seeking terrorist episodes that pose difficult moral and legal dilemmas.

For Americans especially, but also for people elsewhere in the world, there had grown up in the last two or three centuries a faith in progress and an expectation that whatever might be the shortcomings of the present, they could be rectified in the future. Today that faith has waned.[327] To understand its waning we should remember this was foreshadowed as early as 1890 by the Census Bureau superintendent's announcement that America no longer had a frontier of settlement. The delayed but cumulative effects of that change were global rather than merely national.[328] The expectation that the future would be better than the present or past was nurtured by the existence for about four centuries of a New World. Opportunistic expansion of people from the Old World into an unexpectedly enlarged habitat helped democratize their political, economic, and religious institutions, and wrought equally significant changes in family mores.[329] But eventually, as this expansion continued, it had to result in invalidation of the sense of limitlessness that had come to be the central premise of people's lives. The once-New World became more filled up with people than Europe had been when Columbus set sail. *The future*, therefore, *is not what it used to be.*[330]

That is a fact of life we have to recognize if we are to understand the desperation of some of our contemporaries. Providing some agency such as the FBI or the CIA with "the legal and scientific tools" for waging "clandestine warfare" against terrorists will hardly alleviate the oppressive sense of lost limitlessness, or prevent recent changes from nurturing predispositions toward expressive violence. Wiretapping is unlikely to make the planet seem less crowded or the future less constrictive.

Potent twentieth century technology has magnified the power of each of us to get in the way of others (and inadvertently interfere with each other's pursuit of happiness). We are now much more geographically mobile than people were when the virgin hemisphere seemed so endless. Moreover, our per capita resource appetites have been enormously enlarged. So we are vastly more competitive in our quest for shares of the world's finite resources, such as oil from the Middle East. Our competitiveness is channeled by such factors as the division of human societies into contrasting categories—"developed" versus "underdeveloped" countries. But to label the nations that are not yet industrialized "*under*developed" is to go beyond simply denoting their comparative poverty. It is to impute to their peoples the aspiration to become "developed." It presupposes that their destiny does include acquisition eventually of resource appetites as prodigal as those now characteristic of *Homo colossus* countries.

As the twentieth century was drawing to a close there was a crescendo of terrorist activity. One possible explanation may be that as world population approached six billion, more and more people (perhaps especially in regions the "advanced" nations stigmatized as "underdeveloped") came to feel haunted by the thought that they were seen as superfluous.

Thus far the total amount of extra mortality inflicted by terrorist ways of declaring "we do matter" has remained less than the natural increase (births minus normal deaths) of Earth's human inhabitants. So, despite the killing, world population continues increasing—as it did through the two World Wars, and even with the Nazi-inflicted Holocaust.

Much of humanity has become *Homo colossus*, and the notion that all people should aspire to becoming *Homo colossus* has spread to the rest of the world. If in fact Earth's human carrying capacity has already been overshot, the relentlessly increasing *carrying capacity deficit* (the ultimate breeder of redundancy anxiety) will inflict, sooner or later, a catastrophic load reduction. At this twenty-first century's close Earth's human population will be not more, but considerably *less* than at the century's beginning. This is the bottleneck century.

Two Human Species in a Bottleneck

Somewhat more than 30,000 years ago, humans experienced a previous bottleneck. There were then two species of humans alive at the same time—*Homo Neanderthalensis* and *Homo sapiens*, but after a number of generations of coexisting there remained only one human species. We may never know how and why the Neanderthals went extinct. Did *Homo sapiens* defeat them in combat, reducing their numbers to a level below a threshold where extinction soon followed? Or did the modern species of humans simply outcompete the other species in competition for the resources available in the Middle East and parts of Europe where they had coexisted? One way or the other, *Homo sapiens* became the sole surviving human species on the planet, and after spreading to nearly every continent and many islands around the world, we became very diversified—first culturally and racially, then much more recently taking up thousands of different occupations. As I have tried to make clear, this extensive division of labor enabled some of us to become "giants."

So today, we confront another competitive relationship between two broad categories of humans. Hundreds of millions, now equipped with technology that enables us to be ravenous users of fossil energy and other nonrenewable resources, and overusers of renewable food and fiber, are the people I have been calling *Homo colossus.* Our technological equipment changes our ecological relation to the world, and in that sense we can be considered a distinct subspecies. But we members of the colossal subspecies share this one habitable planet with other people who remain less technologically equipped and hence less resource-voracious. Call them *Homo sapiens* (nc)—the non-colossal subspecies.

In an earlier chapter I noted that our cousins the chimpanzee and bonobo are threatened not by what they may be doing to their habitat but by encroachments upon and destruction of that habitat by humans. Similarly, today *Homo sapiens* (nc) is competitively threatened by habitat damage happening globally and attributable not just to their own actions but to the actions of the colossal subspecies.

In the 21st century, a serious carrying capacity deficit puts the two subspecies of *Homo sapiens*—colossal and non-colossal—into an increasingly competitive struggle for existence. This century has become the bottleneck century for humanity because we are an overload beyond the sustainable capacity of the biosphere to provide human sustenance. The load imposed by *Homo colossus* diminishes the life chances both for our own kind and for non-colossal *Homo sapiens* (nc). Insistent pursuit of modern (i.e., colossal) lifestyles will continue the habitat-damaging impacts that will threaten both colossal and noncolossal branches of humanity—and those simian cousins, too, despite their ecological modesty.

To understand it, we should pay attention to a significant preview. Here, among events already happening, are some indicators of experiences humanity will endure in passing through this new bottleneck:

> Major cities in North and South America, Asia and Australia are nearly out of water due to massive droughts and melting glaciers. Desperate farmers are losing their livelihoods. Peoples in the frozen Arctic and on low-lying Pacific islands are planning evacuations of places they have long called home. Unprecedented wildfires have forced a half million people from their homes in one country and caused a national emergency that almost brought down the government in another. Climate refugees have migrated into areas already inhabited by people with different cultures, religions, and traditions, increasing the potential for conflict. Stronger storms in

the Pacific and Atlantic have threatened whole cities. Millions have been displaced by massive flooding in South Asia, Mexico, and 18 countries in Africa. As temperature extremes have increased, tens of thousands have lost their lives. We are recklessly burning and clearing our forests and driving more and more species into extinction. The very web of life on which we depend is being ripped and frayed.

Nobody intended these calamities (as described in this excerpt from a speech by Al Gore in Oslo, Norway, December 10, 2007). They have resulted from demographic increase and the twentieth century's impressive achievements of *Homo colossus*. They are affecting both of today's subspecies—*Homo colossus* and *Homo sapiens* (nc). They reflect damage done to the *global* habitat of both.

We have entered the bottleneck century with insufficient recognition of the real connections among its characteristic woes. Governments, politicians, the news media, and all too many people have been preoccupied with blaming and squelching supposed culprits—from evil dictators to oil price speculators. This is cultural lag. The bottleneck troubles are instead largely an instance of what I have suggested could be called "anthropogenic 'fate'"—the kind of "fate" described by C. Wright Mills. Not simply due to diabolical deeds by evil men, today's most vexing troubles are the unintended cumulative effects of innumerable individual decisions and actions that transformed so many of us into *Homo colossus*—and converted carrying capacity surplus into carrying capacity deficit.

CHAPTER 16

Informative Metaphors

"This country should . . . never act . . . without regard to the essentials of genuine morality—a morality considering the interests of future generations as well as of the present generation."

—Theodore Roosevelt, *Outlook*,
Sept. 23, 1914.

". . . climate change is the biggest problem our civilization has ever had to face up to in its 12,000 years, because it requires a collective response."

—Sir David King, UK Science Advisor, *Science*, 21
Dec. 2007, p. 1863.

"Metaphor is essential. Only it can tunnel through the cortex and jostle our emotional core."

—Connie Barlow, *Green Space, Green Time*. 1997,
p. 34.

Years ago, economists wrote about nations attaining "take-off" into sustained economic growth.[331] Their use of the aviation metaphor seemed to make sense, but only because their discipline typically ignores resource limits—as if a plane's engines never require fuel, or the fuel tank capacity was unlimited (or supplies depend only on monetary prices). Today, wiser heads must recognize resource limits. Even the Earth as our ultimate fuel tank has only finite petroleum content. So time has run out for that metaphor's earlier meaning. Now if we use the "take-off" metaphor it must stand not for launching into perpetual *growth*, but for launching national and global policies of *sustainable* sufficiency. Time is running short for even that more ecologically realistic aspiration.

Enormous social change will necessarily happen as the availability of petroleum inexorably diminishes following peak oil, whether that peak has already occurred or is yet a few years off in the future.[332] Recently there has been extensive discussion about alternative energy sources (alternative fuels, as well as non-fuel energy—wind, solar, etc.), and there is also some attention devoted to conservation, and to ways of doing more with less.[333] Will any or all of these alternatives suffice to overcome our current difficulties?

The fatal crash of Comair Flight 5191 on the night of August 27, 2006 can serve as an illuminating metaphor for the likely fate of our present civilization as time sweeps us beyond the era of cheap and abundant fossil energy.[334] Because of confusion about closure for repairs of part of the taxiway, that airplane made a wrong turn as it taxied out for takeoff from Blue Grass Airport in Kentucky.[335] As they gathered speed that night for getting their prosthetic aluminum wings and jet engine muscles into the air, the crew did not know that they had turned onto a runway that was too short for that particular aircraft. We can consider possible responses to that fateful situation from three points of view: (1) the cockpit crew, unaware of the fatal error until too late, (2) seasoned passengers, with faith in the cockpit crew and in the whole commercial aviation system (3) an alert controller in the tower. Each viewpoint serves as metaphor for different perspectives on civilization's future in the aftermath of peak oil and in an era of global climate change.

When thinking about global climate change, its human causation, and its ultimate impact on human societies,[336] various well-informed people have come to believe that in this industrial era we have *already overshot* sustainable human carrying capacity. My own studies, my travel experiences, and my observations of nature have convinced me that this diagnosis is valid and so is the implied prognosis. Human civilization appears therefore in a situation analogous to that airliner attempting to take off from a runway too short for it to accelerate sufficiently to get airborne before the pavement's end. Carrying capacity deficit is serious, and tantamount to insufficient runway length.

If there is one thing that would be more ultimately disastrous than the loss of personal significance that is now provoking acts of desperation, it is loss of our species' required habitat.[337] We *Homo colossus* are "devouring our own habitat." So we have shortened the runway ahead of us without recognizing what we were doing. We continued *using up* our habitat, in effect—by denying or ignoring limits, and reducing the human carrying capacity of our planetary home by our use of it. The nature of the biosphere upon which our lives depend was being changed by unintended but inexorable consequences of modern resource-consumption, by end-product disposal, and by our myriad other conflicting activities.

Fossil energy empowered our societies to use even Earth's various renewable resources faster than their rates of renewal, and Earth's nonrenewable resources are being depleted by a human population continuing to increase beyond global carrying capacity. The atmosphere we need for breathing is now used as a dumping ground for combustion products generated by *Homo colossus*. From the "fate" we have inadvertently wrought will there be a *magic* escape? Would that be the result of the election by American voters in 2008 of a new federal administration dedicated to restoring "hope"? Could hope's *basis* be restored?

Consider Three Perspectives:

Despite an upbringing that gave me an optimistic temperment, I have never been willing to deny facts, nor to ignore best available ideas for discussing them. So let us view the human situation according to those three perspectives from Flight 5191.

Perspective 1—As we entered the 21st century, defenders of a "business as usual" view of humanity's dependence on petroleum, were thinking like the cockpit crew of the Comair jet must have been thinking as they taxied that night onto the wrong runway for take-off—unaware of the fact that it was going to prove too-short to allow their craft to get airborne. That business-as-usual stance has become a disastrous perspective. The flight crew, after pushing the throttle levers all the way forward as usual to get airborne as expeditiously as possible (never imagining the runway would not suffice), must have discovered the error of their ways only when they were shocked to see under the plane's white lights the runway's end suddenly approaching when they were still seriously short of lift-off speed. The popular media and most politicians who assert a more-of-the-same view of civilization's use of the Earth, naively rejoicing whenever gasoline *prices* come down somewhat, are doomed to experience at some point in the future the same horrified sense of "What have we done!"—likely the last conscious thought of 5191's pilot and co-pilot.

Perspective 2—Energy alternatives (fuels other than petroleum) have been touted in the post-peak era. This could be analogous to an eager exchange of anticipations among the plane's passengers about plans for the next day's activity in their destination cities. Notions that biofuels would enable us to continue high-energy lifestyles unabated even as petroleum supplies dwindle may be just as irrelevant. They run afoul of reality, as succinctly indicated by Richard Heinberg in, *The Oil Depletion Protocol*, where he says

"... trying to replace a substantial fraction of our 85 million barrels per day of global oil consumption with biofuels could potentially overwhelm agricultural systems already destroying topsoil and drawing down ancient aquifers unsustainably." The suddenly fashionable enthusiasm for ethanol as fuel will drive up corn prices and turn our machines into competitors with our stomachs. As a makeshift solution to energy shortages it is anyway an example of "too little, too late."

Perspective 3—The third perspective is reflected in efforts of some of the most concerned and informed people today who are implicitly conceiving their task as persuading human societies to redirect themselves onto a correct path—a path very different from our accustomed ways. One of the best of these efforts is *New York Times* columnist Thomas Friedman's 2008 book *Hot, Flat, and Crowded*, calling for an all-out "green revolution" to reverse the dangerous human alteration of planet Earth. Although he understates the seriousness of our predicament by merely calling the world "crowded" rather than outright *overloaded*, Friedman's effort is analogous to an attempt that *might* have been made by a frantic controller in the tower. The controller would have reacted urgently had he not been so overworked that he didn't see that Comair plane turn off the taxiway onto the wrong runway. The air crew needed to be alerted by radio to the fact that they should *brake and reverse thrust* immediately rather than continuing to accelerate! But it was dark, and that controller had too many other routine matters to attend to.

Prognosis for Humanity

All the previous chapters have been aimed at enabling the reader to see why, with great reluctance and regret, I am compelled to doubt that we can confidently hope to avoid a serious "crash" as the focal human experience of the 21st century—envisioned also as our species having to pass through an ecological "bottleneck."

A few analysts, perhaps most of them in academic or other seriously intellectual jobs, will perceive according to Perspective 3 the challenge now confronting humanity.

Many more, embracing Perspective 2, will imagine themselves realists and will invest their hopes in policies and procedures that will be woefully inadequate as well as difficult to establish, even though they will have an aura of enlightened responsibility, made plausible only by that conventional faith in breakthroughs.

Far too many others, probably a majority of the speaking descendants of those hopeful hominids whose tool-making hands led to enhanced brains that brought us to our present glories and to the impending impasse, will adhere to Perspective 1. That adherence (like the once-standard medical panacea, bloodletting) will enormously worsen our fate. Self-interested perseverance in accustomed ways will be favored by the ubiquitous pickpocket mentality.

Our Need for Knowledge

In the late 1940s, before the pickpocket mentality had become quite so rampant, the peace following World War II seemed already precarious. A nuclear arms race was about to commence, and there was anxiety that a possible Third World War some years hence would involve detonating massively destructive weapons. At that time I read as an assignment in a research methods course at Oberlin College a newly published book that used a question for its title—*Can Science Save Us?*

The answer, according to the book's author, George A. Lundberg, was "Yes, but we must not expect physical science to solve social problems." Good intentions, he said, could not suffice to achieve human goals, either physical or social, if those good intentions remain unconnected to objective knowledge of effective techniques. In a half-dozen chapters his book argued for scientific method being as truly applicable to investigating principles of human behavior as to the study of patterned behavior of planets, atoms, animals, and plants. Persuaded, I acquainted myself with some of Lundberg's earlier publications and then changed my major and my professional aspirations, enrolled in graduate school at the university where he was teaching, bent on acquiring PhD skills that I hoped might yield scientific findings helpful to a quest—shared by many of my peers—for human well-being and world peace.

Through subsequent decades, though, the more I studied the writings of social scientists, even those most ardently attempting to be as scientific as they could in their pursuit of understanding about human societal phenomena, the more I found them too insulated from *other* sciences. It was as if what biologists, chemists, paleontologists, geologists, climatologists had learned and were currently learning about the world we humans inhabit was deemed to have no relevance to predicting future human experience and activity.

So I came to wish Lundberg had also said, "We must not expect even the most explicitly *social* science to 'save us' if it encourages us to remain uninformed about the causes and consequences of events and conditions in the biophysical world in which we live." On the basis of some other things he wrote in the years I knew him, I have no doubt he would insist, were he living today, that as we became hyper numerous and devastatingly colossal in our interactions with this planet substantial awareness of findings in *all* the sciences was growing ever more essential.

Facts of Life

There is another aspect of our predicament we are reluctant to face up to. In addition to consideration of adequate runway length for taking off (even after we have perceptively redefined and updated what we mean by "take-off"), wouldn't it be equally essential to ensure that the "aircraft" (human civilization) was not overloaded. Even on the longer runway (analogue of hoped-for availability of alternative energy sources to power modern societies), an *overloaded* aircraft would need to attain a higher speed to get airborne and not crash. The six-plus billion people aboard this planet as we began the 21st century were and are an overload, especially since so many of us are not simply *Homo sapiens* but through technological progress and extensive division of labor have become *Homo colossus*.

As we pass through the bottleneck century our numbers will perforce diminish. It will be agreed in most circles that a declining birth rate is preferable to a rising death rate as the basis for that reduction in numbers, but there will continue to be sharp differences in attitudes about acceptable versus unacceptable ways by which birth rates might be lowered, by how much they will need to come down, and how soon or how fast. Deep religious convictions will be put to the test by 21st-century conditions. Pressures of political-ethnic competition will contribute to the difficulty, as no interest group will willingly sacrifice the power it derives from its numbers. It is almost certain, therefore, that human numbers will not decline far enough nor fast enough by the preferable means—diminshing reproduction rather than increasing mortality.

To avoid or minimize an actual population die-off, such as befalls other species when they overshoot their habitat's carrying capacity, we must also drastically downsize our per capita "ecological footprint"—reduce our per capita resource appetites and per capita emissions of end-products.[338] No one should imagine any of this will be easy to accomplish.

It will be essential to recognize, acknowledge, and do our best to cope with some enormous impediments to necessary change. In coming to grips with the concept of sufficiency, it may be more important to consider ways of overcoming the seductive effects of two centuries' commitment to industrialism and growth than simply to repudiate the ecological naiveté of contemporary politicians and business executives. But there are a number of reasons why we are unlikely to do so. Among them is the fact that evolution has always been a process by which populations of organisms, confronted with *present* selection pressures, become adapted to *existing* circumstances. If the conditions of their environment change, as in time they must, changed adaptations will be required. That is the story of the origin (and displacement) of species—including human species.

Knowing Why We Exist

If it is true that remaining ignorant of history condemns us to repeat it, then knowing why we exist and why we live as we do may yet be helpful in understanding our future. Some readers may consider asking "Why do we exist?" as equivalent to asking "To what end or purpose are we here?" But there is no scientific way to answer that. Instead, I intend the question to mean "What enabled us to be here?" and science *has* provided knowledge about that. We can go farther and infer from findings of paleontology, ecology, biochemistry, anthropology and other sciences some of the influences that *caused* us to be here and to behave and think as we do.

First of all, human beings, no less truly than other creatures, live by interacting with the other organisms and substances that make up the biosphere. As is true for populations of other species, our effects upon our surroundings affect that environment's subsequent effects upon us. We never could live on this planet without having its nature shape our lives. But the more our activities change the Earth, the more this changes the opportunities and perils with which it confronts us. Human beings are able to be here because prior occupants of this planet (especially single-celled creatures) long ago did things to Earth and its atmosphere in the process of living that happened to make the world eventually less suitable for their own lives to continue, but thereby gave it properties in which other species ancestral to us could arise. That fact needs to become common knowledge.

Our species arose because of selection pressures that acted upon members of a primate species that was like us in important ways but still unlike us in some ways. Earlier hominids lacked some traits that make us the humans we now are. We are a species with traits that have enabled *us* to behave in ways that are changing the nature of this planet upon which our lives depend.

Why are we here? Biologically, as individuals we are here because our parents produced us, after their parents had produced them, etc., etc. Moreover, we are *able* to be here because the planet had the means to support us. We have been able to survive to the point where we could read this book and contemplate our situation because our conspecifics, our fellow humans, inordinately numerous as they became in the past century, have *not yet* crowded us out.

Imagine a species population somehow exempt from dying (having a zero death rate) but with new recruits being added (by a non-zero birth rate). It should be mathematically obvious that such a population would eventually put its descendants in serious difficulty from becoming too numerous for their environment's carrying capacity. No such immortal species exists, of course. By being mortal, people yield their niches to posterity. Among actual (mortal) species, when the reproduction rate *exceeds a replacement level*, carrying capacity limits can cause attrition. The attrition will be not altogether random. That was what Darwin (and Wallace) saw, and they astutely recognized what was implied for species formation by the non-randomness of the inevitable attrition—advantageous traits would be "selected" naturally by enhanced longevity and differential reproductive success.

Why didn't this century's bottleneck hit *Homo sapiens* long ago? In the latter half of the 20th century, general awareness of "the population explosion" caused a popular rumor to circulate that there might be more humans alive in our time than the sum of all past human populations. Careful calculations showed the rumor was far off the mark—and yet there was important truth to be derived from thinking about it.[339]

Even if our genes did not predispose us to something like the proverbial three-score-and-ten life span, the carrying capacity limits of Earth—and of each of its local regions—were real and had real effects. Predation (by microbes as well as by larger predators) was also real.

But now, if we consider Watt's 1776 development of the steam engine, which launched us into escalating dependence on fossil energy, as marking the origin of subspecies *Homo colossus*,[340] it may just about be true that the number of *us* now alive exceeds the total of all prior generations of *this type* of human. Only some of the world's people have become colossal, and only in recent times, mostly within the twentieth century—acquiring per capita power equivalent to scores of "energy slaves."

Cultural evolution, rather than old-fashioned biological evolution, produced *Homo colossus*. We became colossal in far fewer generations than it took hominids to become "fully human." Becoming colossal brought us into the bottleneck.

Evolution and Succession

There are larger contexts, too, in which to consider the matter of living entities making way for subsequent life. Not just individual organisms, or even whole populations, but whole *associations* of interacting populations are involved in intergenerational competition—between each generation and its descendants, because all species and their biotic communities depend on the finite opportunities provided by a finite planet or by its local places.

Every ecosystem consists of an environment and an association of species populations interacting with that environment's physical and chemical characteristics and with one another, taking substances from that environment, reshaping and recombining them by the life-sustaining processes of metabolism, and discarding the used substances in their altered forms back into that environment. The discards from some species may serve as resources for other species, but accumulated effluents tend to be toxic to at least the particular species producing them. The physical and chemical characteristics of an environment are changed by effluent accumulation as well as by resource depletion. Existing niches are diminished and different niches arise. Accordingly, selection pressures change. Some formerly thriving species become hard pressed. Other species, previously ill-adapted to the prior conditions, may now find the changed environment more suitable for *their* way of living and will begin to increase.

When there is at hand a "seed source" for these other species—nearby populations that can promptly enter the changed local environment and flourish therein, the change that occurs in species composition of the local association is what ecologists call *succession*. Species composition of an ecosystem can change in a few generations by the process of succession, involving local addition of already existing species formerly living elsewhere but within accessible proximity to the newly suitable region.

Consider ecosystems dominated by humans. Ecosystems in the western hemisphere became drastically more human-dominated as a result of invasion by Europeans. The technology that converted many of the people in both hemispheres into *Homo colossus* wrought tremendous ecosystem modification everywhere.

When the life of a biotic community changes its environment in ways deleterious to some of its constituent species, and there exist as yet no "replacements" that would be adapted for the changed conditions, succession may be stymied. But the more stretched out process of *evolution* may ultimately provide the successors. Evolution draws upon the variation among individual members of an existing population, with statistical differences occurring in their reproductive success. Those variants which happen to be more nearly fitted to the changing environmental conditions increase in number. Eventually this process of natural selection, by producing *new* species to fill new niches, may accomplish ecosystem change similar to ordinary succession.

What matters to us in the 21st century is that in both the shorter time perspective of succession and the longer-term perspective of evolution, *elimination of species populations that are unable to adapt to changed environmental circumstances will be a natural occurrence.* As life for those species becomes more difficult, reproduction falls below replacement rate. It may take many generations with sub-replacement reproduction to reach extinction, or sometimes the result may occur more swiftly. In general, members of endangered species will only know their lives are difficult. Nonhuman creatures going through a bottleneck or facing extinction will not know either the bottleneck concept or the extinction concept. Ignorance of the pertinent concepts is no protection from the perils they represent.

Are the special qualities of the human species reasons to hope for exemption from either of those experiences? In general, species are not typically exempt from either bottlenecks or extinction. Nonhuman species are generally unable to foresee the bottlenecks ahead and the extinction risks that their own ways of living may engender. Humans, with language, *can* know about bottlenecks and about extinction. For humans, neither ignorance nor denial will suffice to prevent a bottleneck's consequences.

The special human traits—including speech and language—which enabled us to diversify culturally and occupationally while remaining biologically one species, did cause human beings to become increasingly interdependent. On that point Durkheim was right, but in the final decade of the 19th century, he was unable to foresee how far this interdependence would have been carried by the beginning of the 21st century. *Homo sapiens'* powers of foresight may be greater than those of pre-human species, but they are not unlimited. Like all other species, because of the very nature of natural selection, we tend to be preoccupied with the here and near.

Global Cooperation Among the Dehumanized?

For there to be any hope of solving the most serious problems confronting the world today, even after we have begun to perceive their connection to the abusive dominance of the biosphere by *Homo colossus*, unprecedented society-wide and world-wide cooperation is urgently needed. Tragically, the insidious identity-denigration tendency so intrinsic to modern societies is a serious impediment to cooperative efforts beyond the local neighborhood or community. And the degradation of *nations'* identities characteristic of international "trading" practices may preclude international cooperation even when it is essential. I am not talking about capitalism versus communism, nor about Christian ideals versus Muslim ideals versus those of any other religion. I am referring to the degrading effects universal to all large-scale societies with industrial-level division of labor.

Within small groups, face-to-face interactions may have kept at bay the pickpocket mentality, but as society after society has become predominantly urban, there has been a burgeoning of oppressive anonymity. As division of labor grew more and more extreme, compounding our interdependence, we all had to look to others as sources of life's necessities and instruments for the satisfaction of our wants. The person-reduced-to-tool status became a core aspect of human experience. Dependence upon a money-mediated exchange network deepened this dehumanization. How could the imposition of indignity by the *significance-deprivation* when people are implicitly reduced to the status of tools to be used, or pockets to be picked, not resemble the effects of slavery or of life-sentence incarceration?[341]

Rather than begetting the organic solidarity so hopefully anticipated by Durkheim and his intellectual heirs, modern division of labor has too widely undermined our significance as human individuals. No wonder there are occurring in our world today such violent "screams"—screams in behavioral forms that range all the way from displaying angry or defiant bumper stickers, surreptitious daubing or spray-painting of outrage-expressing graffiti, or mere shouting of obscenities, to riotous uprisings, to suicide bombing of innocent crowds—screams declaring in desperation, "We exist! We are human; we count for something—even as we die!" Do other people, following the news in a world of fully developed diversity, hear these cries, or do they usually just cringe at the thought of those horrid "others" who seem addicted to misbehavior?

It is not just militants, rioters, or terrorists who feel significance deprivation, though their actions may be more overtly expressive of that loss than the sense of desperation suffered covertly by many others. If even in the best of times, members of modern societies with industrial-level division of labor can feel *used*—dehumanized in varying degrees by their consignment to the status of *tools*, or by the pickpocket culture's definition of them as revenue-source *customers*—then when the economies of nations fall into "hard times" and growing numbers of human tools are "laid off" they are thereby required to endure an ultimate insult, a last-straw diminution of personal worth essentially to zero. When families are uprooted by losing their homes in a crescendo of mortgage foreclosures, their members lose personal dignity as well as material shelter.

Violent outbursts will not be stopped by righteous declarations that the screamers are fanatics, or mentally deranged. A surge of protest behavior around the world implies there has been a pandemic of dehumanizing experiences. Militants, whatever their ostensible ideology, whatever their particular complaints and demands, declare by their actions, "My existence does matter! I can make a difference. See, I *am* making a difference!" (Or, posthumously, "I *did* make a difference!")

Can we really imagine there is any way of ending terrorism and other cries of outrage without having to undo the social arrangements that inflict this now pervasive significance-deprivation?

Into the Bottleneck Century Hobbled by Our Pickpocket Culture

But *is it possible* for the identity-diminishing social arrangements evolved since the industrial revolution ever to be undone? Having become both numerous and voracious, haven't we humans (at least we of the *Homo colossus* subspecies) become utterly dependent on fantastically elaborate division of labor, with all those dehumanizing effects of the necessitated exchange relations?

What hope is there for stopping our self-destructive *Homo colossus* habits?[342] Can it be done soon enough, and universally enough to retrieve our future?

We could begin to hope if and when an American president refrained from ritualistically repeating early in his annual address to Congress the applause-line: "The state of our union is strong"—and honestly declared instead that it is *precarious* because of what human action has done to the biosphere.

When Norway's Nobel committee awarded half of the 2007 Peace Prize to the Intergovernmental Panel on Climate Change (a network of 2,000 scientists who amassed an informed consensus about the connection between human activity and global warming) and the other half to the "single individual who has done most to create greater worldwide understanding" of what must be done, those Norwegians were telling all of *Homo sapiens* something vitally important. It was something that the subspecies *Homo colossus* did not want to hear. They were telling us that the IPCC's measurement of the runway had found it terribly short and growing rapidly shorter. To members of that Nobel Committee it was clear that Al Gore (having been excluded seven years before by the U.S. Supreme Court from the presidential role) had taken on the role of alert controller in the tower. The Nobel committee in effect declared he was telling *Homo colossus* "Wrong runway!"—trying to arouse us to see that we must brake rather than accelerate, and change course immediately.

What Now?

If we had really got started *right then* trying to become drastically less colossal in our ways, weren't we already starting too late? After all that we humans did to this planet and ourselves in the 20th century and before, only extraordinarily fundamental change in our ways of living could save us from devastating pressures in the 21st century bottleneck. Change had to be swift and global. Unprecedented levels of interethnic, interfaith, and international cooperation were necessary to counteract dire trends so utterly translocal as climate change, resource depletion, and ecosystem disruption.

But human shortsightedness persists—because it had resulted originally from the "adapt-to-*present*-circumstances" nature of the evolutionary process—and it became seriously aggravated in recent decades by prevalence of dehumanizing influences (epitomized in the pickpocket perspective produced by modern division of labor). Human inability to relate wisely to even a just beyond-the-horizon future had thus appreciably diminished the prospect for ever achieving truly unselfish cooperation on anywhere near the required scale.

Though we moderns may be stubbornly ethnocentric about our "right" to plunge onward with "our way of life," no social order systematically promoting the pickpocket outlook and the impatience that insists we must "fly now, pay later" can long endure. Our descendants may see that in retrospect.

Epilogue

I rejoice at having lived in the time just preceding the bottleneck century, even though the view ahead, as these 16 chapters have enabled us to discern it, must appear extremely disheartening. Genuine knowledge, however mixed a blessing it may sometimes seem, is never as oppressive as languishing in bewildering ignorance.

Mine has been an exciting life, spanning the last three quarters of the twentieth century and into the first quarter of the twenty-first. In no comparable span of previous time has so much new knowledge arrived, nor so much human history happened. What has been learned enables us to understand, among other marvels, how *Homo sapiens* came to exist and how we came to behave as we do. Knowledge of this planet and the living beings it supports and of the galaxy and universe in which it exists has increased "astronomically" within my life span.

There are now known to be more than just the two broad organic kingdoms—plants and animals—recognized in Darwin's time. We have learned how our planet has drastically changed from its original form, how it was fortuitously rendered suitable for animal life by the biogeochemical effects of preceding microorganisms that were cumulatively spoiling it for their own kind. It is now understood that eventually a warm-blooded portion of kingdom *animalia* became an order we have come to know as primates. That order included apes, some eventual descendants of which were bipedal hominids, a few of whose still later descendants developed such dexterous hands that the selection pressures affecting them caused some of their descendants to be larger-brained, symbol-wielding humans—with rapidly evolving technology and ever-changing diverse cultures.

The burgeoning technology and the social differentiation enabled by those cultures would lead to such extensive division of labor that self-sufficiency would vanish, and would necessitate money-mediated exchange networks tending to engender pandemic self-centeredness which would seriously impede rational cooperation required to cope with changing future circumstances.

But at least some of the events that occurred during my lifetime ensure that *Homo sapiens* was, even so, a marvelous achievement of the evolutionary process. The handsome face of a high school friend of mine had been severely scarred by smallpox, but forty years later, after a worldwide effort, that disease was declared eliminated from the world. My parents once rerouted a family vacation itinerary due to worries about cases of contagious polio reported in a region through which we had intended to travel. By the time my wife and I had progeny to protect from the scourge of that crippling disease it could be done by vaccination.

Some of the achievements of *Homo sapiens* converted a portion of humanity into *Homo colossus*. In retrospect, we can regard that transformation as a natural product of the largely myopic evolutionary process, propelled by adaptation to existing (not future) conditions.

In 1945 I was privileged to experience the *thrill* of being colossal. Following the capitulation of Imperial Japan at the end of the Second World War, the U.S. Navy's entire Third Fleet assembled in close enough proximity for a victory photograph. We had been operating in the western Pacific through the war's final months as widely separated task groups of some two dozen ships each, every group out of sight of all the other groups, the totality having remained essentially *unknown* to any personnel much below the rank of admiral. The large aircraft carrier which had been my home for more than a year was suddenly in the midst of other gray vessels from horizon to horizon, some shaped just like us, others very different. They were so seemingly innumerable that scanning the scene was as humbling yet spine-tingling as counting stars in a clear night sky. What an awesome display of industrial output it was, by a proud nation then less than half as populous as it would become by that century's end. But my education was not then far enough advanced for that sight, albeit breathtakingly impressive, to lead me to coin the term *Homo colossus* until several decades later.

Born the year before Charles Lindbergh's solo flight from New York to Paris amazed the world, I was amazed in my teens by a *National Geographic* photograph taken by men on a record balloon flight that had reached an altitude high enough that the picture actually showed the horizon's curvature! I lived to see photographs of the whole round Earth from space, and on into a time when people had already become indifferent about men having gone to the moon and returned home with specimen moon rocks.

My generation grew up not knowing about "the ozone layer" protecting us from ultraviolet health-hazards, and though we learned in school that the world had once had much more extensive glaciation than the polar ice caps of our own era, we were not anticipating the further deglaciation we've seen since, nor had we any comprehension of the many biological, agricultural, and societal repercussions of climate change now beginning to manifest themselves.

In short, I was privileged to live through the best years of the reign of *Homo colossus*—a time when the benefits of being colossal were maximized, and before we were overtaken by their ecologically disastrous denouement. Too late, we have begun to see the rapid approach of our runway's end. I have pity for all who insist "there's still time." I deplore those who naïvely count on merely "stopping the clock" by some yet-to-be-made miraculous breakthrough.

Of course I hope my little great-grandsons (there are now two of them) may be among the survivors of that passage through the twenty-first century bottleneck my generation and our ancestors have made inevitable. I hope they will be enthusiastic in learning about many interesting topics. Despite the hardships occurring in their lifetime, I hope they will retain a sense of wonder and an ability to marvel at the world they inhabit and to appreciate life's many joys, big and small, past, present, and future.

Moreover, I hope by the time they become great-grandfathers themselves, their generation will be so conspicuously more enlightened than mine was and our forebears were that the world population of bottleneck survivors will have evolved social systems better able to be circumspect in their use of their planet and its vulnerable biosphere. If readers of this book come to share similar hopes, and contribute to instilling them in their descendants, my reasons for writing will have been justified.

ENDNOTES

Chapter 1

1   See Blau 1977, pp. 6-11.

2   Durkheim 1984 [1893].

3   See, for example, Gunther 1950, p. 278; Gallagher 1985, p. 97. Meyers 2006, p. x, suggests that "When FDR said, 'The only thing we have to fear is fear itself,' he could just as well have been inscribing the epitaph on a postmodern American tombstone. Now, it seems, the only thing we have to fear more than terrorism is fear of terrorism itself."

4   Rumsfeld 2006.

5   Zernike 2006. Perhaps because of the aging of our population, over a 17 year period, NYPD statistics reflected an on-going decline in urban crime, with murders in New York having declined to as "few" as 537 in 2005, as compared with 2,245 in 1990. See Baker 2005., and Szep 2006.

6   Thompson 2006.

7   As one observer put it, "The violent movies that we had stopped making for a few months came back, and once again we cheered mindless death and destruction." See Meyers 2006, p. x.

8   On the commercial exploitation of this fear, see, for example, Pillar 2006. The infection has crossed from birds to other species, including, in Germany both a cat and a weasel-like animal called a stone marten. See Eddy 2006, and McHugh 2006. An associate director of the Center for Health and the Global Environment at the Harvard Medical School wrote to *The New York Times* to point out that this avian flu was already "a global pandemic" among birds.

9   See Cortazal 2006, and Connolly 2006. The prospect for widespread human infection by the bird flu virus seriously alarms one United Nations expert. See McNeil 2006.

10   Siegel 2006.

11 Pan 2006; see also Efron 2006.
12 Hafner and Rai 2006.
13 Kurtz 2006.
14 Dalton 2006.
15 See BBC News 2006; and "Economic and human costs of the *Jyllands-Posten* Muhammad cartoons controversy," from Wikipedia, the free encyclopedia (on the Internet). A Lebanese-born professor at Sarah Lawrence College (Gerges 2006:238) described the reaction as "an expression of an identity crisis—a hypersensitive state quick to perceive internal or external danger. According to most public-opinion surveys, ordinary Muslims as well as Islamists already fear that the 'Christian West' is waging a war against them, one in which their spiritual values—their very Islamic identity—are being targeted. The Mohammed cartoons fed straight into that fear."
16 Antipathy can be generated in many other ways than by a nation's official foreign policy. People in other lands are affected today by American activities, political, economic, religious, and otherwise—just as we are, often to our disbelief, affected by the activities of various parts of the 95 percent of humanity who are not Americans. Many of the word maps we generate in America about ourselves and about others are beheld abroad, along with other images (e.g., Hollywood films and U.S.-made TV programs) reflecting our conceptions of what being an American involves and what it means to be other than American. It should come as no surprise that we are less than universally loved and are often resented. See Sardar and Davies, 2002.
17 For clarification of what is properly meant, and what is not meant, by saying we were "destined" to produce these things, see Chapter 9.
18 See Durkheim 1984 [1893].
19 See Ross 1973 [1907].
20 Roughly half of the world's people now live in cities. For most of our species' existence, nearly all human beings lived in small aggregations, profoundly unlike today's cities. Even in the centuries of written history most people dwelt in rural areas, villages, or small towns, only a very small minority of even our recent ancestors were urban dwellers. Human attention spans are limited, so the number of people we can know and relate to personally is far less than the number clustered together in today's large "urban agglomerations." Substantial anonymity is an ineluctable fact of urban living, and alienation easily flows from it. The aggregation of humans into swollen cities reflects not only the growth of population but also the division of labor, as the interdependence among diverse occupational specialties exerts a pressure for proximity to each other. Whatever product or service one has to sell, proximity to the "market" for it is sought. It is thus that the vernacular usage (and the sociological usage) of the term "community" to refer to a town or city is consistent with the ecologist's usage of the term in reference to an association of interdependent species populations in a limited geographic area. Differently specialized humans function

analogously to different species even though we are all one species biologically. The term quasi-speciation to refer to occupational differentiation made such sense to me (see Catton 1980) I began using the phrase before I knew of its prior use in ecological-evolutionary theory by Hutchinson (1965).

21 Ethnocentrism manifests itself in many ways. The desire for national heroes (and even national) saints is an expectable expression of this human tendency. For example, see Wilkinson and McDonnell, 2007.

Chapter 2

22 For an insightful analysis and elaboration of this "each in its respective sphere" idea, and a cogent argument for humanity's need of both, see Gould 1999.

23 See Love 1956 for the story of King's presidency.

24 These "libe dates" may sound like necessarily prosaic experiences, but in retrospect, one can regard such occasions as combining the "best of both worlds"—companionship with a member of the opposite sex and acquisition of knowledge, two goals of attendance at a coeducational college or university. Seating patterns in the main reading room commonly reflected this. Seldom were a table's six chairs all occupied by students of just one sex. Often there were three males and three females, not infrequently in alternate seats. If one were to glance under such a table, occasionally one might see between adjacent chairs an intertwining of male ankle and female ankle.

25 Since this was long before the beginning of the jet age, which familiarized people with delta-wing aircraft, it is perhaps forgivable that my teacher "misconstrued" my aeronautic intention. In my teens I used my hands to build ship models and model airplanes. Many years later, my own fondly-remembered model-building experience with my two hands caused me to feel a sense of utter amazement at the dexterity and personal achievement by a one-armed colleague in the university where I was teaching, when he showed me his spare time construction: an exquisite rubber-band-powered balsa-and-tissue flying model airplane. Just as inspiring are the occasional true stories of men actually playing baseball despite loss of an arm (e.g. Pete Gray, outfielder with the 1945 St. Louis Browns), or being born without a right hand, such as Jim Abbott, who actually pitched for several major league teams, even throwing a no-hitter for the NY Yankees against Cleveland in 1993).

26 The book was originally published in 1904 and was intended for educational use in opening new knowledge vistas for young children. When I looked it up again many years later, after having earned a PhD in sociology, and having taught in a number of universities, I was pleased to find that one of the helpful critics acknowledged by the author was a University of Chicago sociologist deservedly famous early in the history of that institution and its already eminent sociology department.

27 Only recently, however, I learned that those saber-tooth felines were already extinct

before the people in that book would have existed. Had the story book's author known that, presumably the joy of my education in the second grade would have been diminished.

28 This was before the taking of enormous boatload after boatload of the sand from those dunes, down the Great Lakes to be used in foundries and glass factories. Opportunities for soaring on updrafts must be more restricted along that portion of the lakeshore for today's gulls. Of course none of them knows about the fabulous soaring opportunities that were available to their ancestors. Soon there will be few human residents of that town with memories of those sand dunes to make them wonder about the wisdom of our species changing the face of the planet that produced us.

29 Bishop Samuel Wilberforce (1805-1873) was not the abolitionist William Wilberforce (1759-1833), but was William's son.

30 Many fossils found in the 20th century, together with modern methods of reliably dating the age of these fossils, have provided conclusive evidence (from the shape of leg bones, pelvis, and location on the skull of the point of its attachment to the spinal column) that we had ancestors who walked upright on two legs many generations before our ancestors had large brains.

31 In a comparison of human upright walking and chimpanzee knuckle-walking, for equivalent energy-expenditure the human goes approximately 50 percent farther.

32 The apes we descended from were not any of the ape species alive (and increasingly endangered) today. Today's apes are our cousins, not our ancestors, but some of them (especially the chimps) may be assumed to resemble substantially the species long since extinct but ancestral both to them and to us. The feet of subsequent hominid generations ceased to be so hand-like. When my youngest grandson was a toddler I showed him a group of apes in a zoo and then taught him to call the human foot "a degenerate hand," much to the amusement of other adults. The big toe was no longer an opposable thumb. Instead of human feet retaining the ape foot's ability to grasp a tree branch, the human foot needed toes shaped and aligned for bearing weight when "stepping off" on the next stride in full time bipedal walking.

33 See Stokoe (2001).

34 These paragraphs describing the early stages of language learning are based largely upon Yang (2006: 66). See also Petitto and Marentette (1991).

35 See, for example, Stokoe (2001: 132-135).

36 This idea is well developed by Lieberman (1998) in connection with the premise that speech is our most essential human attribute.

37 Note that in some dialects (e.g., social class variants of English) the f or v sounds actually displace the unvoiced and voiced th sound. Some people routinely say "wiff" for with, and "uvver" for other. For a more extensive discussion of such exploration, see Yang (2006:61-63)

[38] It was by carefully watching my lips and the tip of my tongue that one of my grandsons, at age 2, learned to say "feather" instead of his prior mispronunciation, "fezzer."

[39] Never again would I boldly imagine first drafts (of my own writing efforts especially) could serve as finished products. It was probably the 1933 reprint (by a different publisher) of Becker's book that I read. The book should continue being influential. It has since been reprinted at least twice more (by still other publishing companies).

[40] For an illuminating compilation and discussion of excerpts from Darwin's notebooks and preliminary essays representing steps in his working out the theory and evidence for it over several decades prior to final publication of the Origin in 1859, see Glick and Kohn (1996). Their book approaches the theory of natural selection in a way that is equivalent to Becker's study of the Declaration of Independence. See also various excellent biographies of Darwin, especially Browne (1996, 2002); Desmond and Moore (1991). For special insight into the personal agony in Darwin's life as a scientific thinker see Keynes (2002) and Quammen (2006). And for a detailed analysis of evolutionary theory as further developed by scientists in almost a century and a half since Darwin completed The Origin, see the huge treatise by Gould (2002).

[41] See Wilford (1990); and Mckie (2000):54-55. For an earlier, but more detailed, account of the entire history of Piltdown Man, from the original "find" through the controversies about its meaning, to the determination that it had been a hoax, see Reader (1981).

[42] Yang (2006:2).

[43] On the importance of the definition of the situation in shaping human action, see Thomas (1923:40).

[44] See, for example, the preamble to the United Nations Charter.

Chapter 3

[45] Thomas (1923):41-43.

[46] See Coser (1977):521.

[47] For documentation of W. I. Thomas's sole authorship of "the Thomas theorem," see Merton (1995).

[48] "If people believe in witches," said Coser (1977:521), "such beliefs have tangible consequences." These may, for example, include killing those persons taken to be witches. The human mind has the power to transmute raw sense data into a potent categorical apparatus. "Once a Vietnamese becomes a 'gook,' or a Black a 'nigger,' or a Jew a 'kike,' that human being has been transmuted through the peculiar alchemy of social definition into a wholly 'other' who is now a target of prejudice

and discrimination, of violence and aggression, and even murder." Coser adds that there can be benevolent as well as malevolent consequences of definitions of the situation.

[49] See, for examples, Anderson (1996): 81; Angier (2005); Diamond (1992): 147-148; 196-199; Haddock, et al. (2005). Also, see "The Common or Virginia Opossum (Didelphis virginiana) . . . When struck it feigns death, lying inert with tongue lolling out and eyes closed." in *The Larouse Encyclopedia of Animal Life*. New York: McGraw-Hill Book Company, 1967. p. 482 And see, Bird-dropping Spiders <http://www.amonline.net.au/factsheets/bird_dropping_spider.htm>

[50] Sellin (1938)

[51] Cf. Sutherland and Cressey (1960).

[52] Cf. Cloward and Ohlin (1960).

[53] Ross (1907):3-5.

[54] See, e.g., Childs and Cater (1954):99-100; see also Magnuson and Carper (1968).

[55] MacKinnon (1978):178

[56] MacKinnon (1978):179-180.

[57] MacKinnon (1978):181.

[58] MacKinnon (1978):185.

[59] Owen (1980):17.

[60] Owen (1980):9.

[61] According to Wickler (1968:186-190), experiments have shown birds do distinguish eggs not closely resembling each other and will eject from the nest an egg it recognizes as unlike the others. Although it is the odd egg, not necessarily a foreign egg, which the avian parent ejects, experimental evidence shows that the birds which become hosts to the parasitizing cuckoo distinguish their own eggs more exactly from those of a stranger than do other species. This is perhaps an adaptation, a protective mechanism, resulting from the pressure of cuckoo parasitism. A bird which can raise only its own young has more descendants than a competitor which is deceived into feeding a cuckoo.

[62] The failure of deception too-often attempted received ancient recognition, as the theme of one of Aesop's Fables—about the shepherd boy who enjoyed laughing at his neighbors who came running when he cried "wolf' with no wolf actually present, only to find his flock devastated when the disgusted neighbors refused to come when his alarm call happened to be true.

[63] Although such oscillation has been found in other parasite-host relationships, no exact experiments known to Wickler (1968:191) had yet been carried out on the cuckoo. There is a fascinating indication of the human relevance of understanding the cuckoo's special form of mimicry. For a book about computer hacking, Stoll (1989) chose *The Cuckoo's Egg* as its title. The author, an astronomer by training,

became by accident a leading authority on computer security, has delivered numerous lectures on the subject, even to audiences at the CIA and NSA, and has appeared before the U.S. Senate. In an episode when he was studying fresh computer printout to monitor the activity of a particularly persistent and troublesome hacker, Stoll (1989:153) wrote, "I watched the cuckoo lay its egg: . . . use the Gnu-Emacs move-mail to substitute his tainted program for the system's atrun file. Five minutes later, shazam! He was system manager. Now I had to watch him carefully. With his illicit privileges, he could destroy my system, either by accident or on purpose."

[64] See Codevilla (1997a): xvi-xvii. But Machiavelli should not be "accused of saying that bloody dishonesty is always the best policy" because he did note that "force and fraud lead as often to disaster as to success".

[65] Of virtuous qualities, he adds: "I will be so daring as to say this, that, having them and observing them always, they are harmful, while appearing to have them is useful; like appearing piteous, faithful, humane, integral, religious" (Machiavelli 1997 [1532]:66-67).

[66] See Grodzins (1949).

[67] Brokaw (1998):216.

[68] Grodzins (1949): 60, 205.

[69] Grodzins (1949):206. On more than one occasion, "required by the military" has been a thought-stopper. To have learned anything of practical significance in the future from *Americans Betrayed*, we must try to recognize whatever other symbols may acquire prestige as numbing as "national security" or "military necessity" became. Whenever such iconic symbols are in vogue, extreme caution becomes essential regarding the sort of procedures, policies, attitudes, and goals which we allow them to persuade us to accept. Mentions of "Military necessity" wrought their mischief again during the war in Iraq; see Lichtblau 2008.

[70] It would be unwise to allow our natural anthropocentrism to suggest that definitions of situations are uniquely human, for that supposition could be the kind that (as the pioneer geologist Charles Lyell so wisely said of prevalent biases in his time) "may render us blind to facts, which are opposed to our prepossessions, or may conceal from us their true impact when we behold them." Lyell was quoted thus by Stephen Jay Gould (1999:90) who expressed the same caution in an essay about the debate over uniformitarianism vs. catastrophism by saying such comprehensive worldviews "provide both joys and sorrows to their scientific supporters: the great benefits of a guide to reasoning and observation, a potential beacon through the tangled complexities and fragmentary character of nature's historical records— ineluctably combined, however, with the inevitable, ever present danger of biases and false assurances that can blind us to contemporary phenomena standing right before our unseeing eyes." Intelligent people (sociologists included) should ask

which way anthropocentrism has turned out: a potential beacon, or a blinding false assurance.

71 This was the theme of one of Aesops' fables, about the shepherd boy who teased his neighbors into running to his aid when he falsely cried "Wolf!"—only to find they no longer came to help when his cry of alarm was real, as wolves actually did attack his flock.

Chapter 4

72 See Catton 1980: 152, 279. The term "quasi-species" had been previously used by G. E. Hutchinson (1965), a biologist, to denote human groups differentiated culturally (and not biologically) from each other.

73 The experience would be very similar if you scanned rather thoroughly the "Help wanted" ads in the classified section of a big city daily newspaper, though there would doubtless be an appreciable fraction of positions available not requiring a high skill level.

74 See, for example, Schoener 1974, Roughgarden 1976, and Pacala and Roughgarden 1982.

75 Darwin 1859. Durkheim's theory about division of labor was contrary to Spencer's in important ways, but it was no less concerned with long-term changes seen in evolutionary terms. Thus it would hardly have occurred to Durkheim not to cite Darwin. And Durkheim also explicitly cited Milne-Edwards as one of the sources of "recent philosophical speculations in biology [which] have finally caused us to realise that the division of labour is a fact of generality that the economists, who were the first to speak of it, had been incapable of suspecting." Durkheim 1984 [1893]:2, my italics. Cell structure has been greatly illuminated by subsequent cytological research that was hardly possible in Milne-Edwards' time, revealing division of labor within cells as well as among cells. It is now understood that "the various parts of the cell perform important but different functions according to the task *for which each is structurally and chemically suited*" (Bourne 1962:v, emphasis added). Neither Spencer nor Durkheim was in a position to consider that level of biological knowledge.

76 Blair 1959:52.

77 Ibid., italics added.

78 Spencer 1862:382.

79 Durkheim 1984 [1893]:206.

80 Ibid.:207.

81 The words of Durkheim seemed to be discounting any specialization-inducing power of "social conditions." Was he neglecting at this point his own discernment of "moral density" as a factor leading to specialization? Even great minds can be inconsistent.

82   For a clear and simple definition of "niche" reflecting modern ecological usage, see Odum (1989:50). For a revealing summary of the way the concept has evolved since it entered the ecological vocabulary, see Schoener (1989).

83   Price 1984:354.

84   Sometimes (but not invariably) these communities comprise species that are more generalized in their exploitation patterns. So the differences in extent of "species packing" in these different regions would appear to support Durkheim's idea about the competition reducing effects of division of labor. However, the species found in temperate regions generally appear to be about as specialized in their feeding habits as those in the tropics, says Price (1984:354). Certain beetles "are actually more specialized in host plant utilization in temperate regions than in the tropics . . . , butterflies seem to be equally specialized along the latitudinal gradient . . . , as do marine monogenean parasites on fishes . . ." The concept of species packing and its relation to competition has been analyzed by biologist Robert MacArthur (1969; 1970).

85   On various aspects of the relation between resources diversity and species diversity, see Johnson and Raven (1970); Valentine (1971); Powell and Taylor (1979); Bonner 1988

86   The assortment of relations between populations of different species includes mutualism, parasitism and predation, and other relations called commensalism, competition and amensalism, plus the condition of neutralism or noninteraction.

87   The actual relative importance of the various kinds of relationships needs further study.

88   It was mutualism as well as specialization Paul was depicting to the Christian converts in Corinth.

89   However, if we were to re-express Durkheim's views in more explicitly ecological and evolutionary terms, these views may turn out to entail important insights about the causes, direction, and consequences of human division of labor that he never suspected. Careful consideration should be given to other aspects of Darwinian evolution, apart from the relation of niche diversification and coexistence of species. Newer knowledge about how symbioses arise, and how they change, will require significant amendment of some of Durkheim's conclusions.

Chapter 5

90   See Catton (1980).

91   The evolutionary biological term speciation means differentiation of a population's descendants into distinct types that do not interbreed and that have genetically-based traits that are adapted to different life circumstances. By speciation, descendants of one species become two or more species. Specialization (within a human labor force) means each worker follows a limited line of endeavor, rather

than being a "jack of all trades." The greater the division of labor, the narrower and more numerous the occupational specialties. As sociologists today clearly recognize, differences between one human specialist and another are not necessarily nor usually genetically based—although, as between, say, a National Football League linebacker and a jockey, their occupationally relevant difference in physical bulk may well have been partly due to parental genetic differences. Opposite sex descendants of different occupational specialists would ordinarily be capable of interbreeding, whatever may be the sociocultural influences toward assortative mating that might keep them from doing so. Fertile offspring may be produced by the mating of an operatic soprano and a baritone, and of course such offspring may well become nurses, airline pilots, lawyers, or bankers, or even sociologists, rather than following careers in music.

[92] See Thompson 1982:61-69, and Rickleffs 1979:226.

[93] See Popper 1968:162.

[94] Students of sociology often were taught to deplore Herbert Spencer's use of "the organic analogy"—a metaphor depicting occupational specialties as "organs" serving different functions in society seen as an "organism." But the idea long antedated Spencer. Among philosophical treatments of division of labor should be included the following Biblical metaphor (1 Corinthians 12:17-28): "If the whole body were an eye, where would be the hearing? If the whole body were an ear, where would be the sense of smell? But as it is, God arranged the organs in the body, each one of them, as he chose. If all were a single organ, where would the body be? As it is, there are many parts, yet one body. The eye cannot say to the hand, 'I have no need of you,' nor again the head to the feet, 'I have no need of you.' On the contrary, the parts of the body which seem to be weaker are indispensable, and those parts of the body which we think less honorable we invest with the greater honor, and our unpresentable parts are treated with greater modesty, which our more presentable parts do not require. But God has so adjusted the body, giving the greater honor to the inferior part, that there may be no discord in the body, but that the members may have the same care for one another. If one member suffers, all suffer together; if one member is honored, all rejoice togther. Now you are the body of Christ and individually members of it. And God has appointed in the church first apostles, second prophets, third teachers, then workers of miracles, then healers, helpers, administrators, speakers in various kinds of tongues."

[95] Smith (1965 [1776]): 259.

[96] Smith (1965): 7.

[97] It became altogether customary to think of machines as "labor-saving" devices, but this neglected the fact that they also fostered "gigantism" among modern human populations—giving us enormous new capabilities and turning Homo sapiens into *Homo colossus*. The fundamental significance of this is emphasized in later chapters.

[98] Spencer, too, passed over the power-enlarging effect of technology and remained preoccupied with heterogeneity-enhancement as the supposedly most sociologically salient effect of technology.

[99] For a lucid summary of what Spencer had to say about the development and importance of division of labor—a summary that clearly reveals significant parallels between Spencer's and Durkheim's views, see Perrin 1976, especially pp. 1349-1353. For an earlier critical appraisal of Spencer's view of the nature of progress, see the Herbert Spencer Lecture by Ensor (1946). Charles Horton Cooley shortly after he had served as the eighth president of the American Sociological Society, said Spencer's *The Study of Sociology* "probably had done more to arouse interest in the subject than any other publication before or since." See Cooley 1920:129, and Spencer 1961[1873]. That book by Spencer had in fact been used at Yale by the American Sociological Society's second president, William Graham Sumner, when he taught in 1876 the first course in sociology offered in any American university.

[100] Spencer (1961): 305.

[101] After studying medicine and zoology in Paris, Milne-Edwards became a French citizen (and wrote in French). His ideas about tissue specialization, or division of labor within an organism, were set forth in 1858 in his *Introduction à la zoologie générale, ou considérations sur les tendences de la nature dans la constitution du règne animal*; see Anthony (1981). Darwin (1859:116) had also cited Milne-Edwards in regard to physiological specialization.

[102] Spencer (1961): 305.

[103] Durkheim (1984 [1893]).

[104] The possibility of inequitable distribution would hardly tend to foster "organic solidarity."

[105] This changed usage can have the effect of sometimes invoking pejorative terms that may impede clear analysis.

[106] That story was in our minds recently as my wife and I, on an automobile trip, approached Hannibal, Missouri, the boyhood home of Samuel Clemens, who wrote under the pen name, Mark Twain. We had both read the story as youngsters, and had both visited Hannibal previously as adults, where we had seen the home and the white-painted fence, but we found ourselves disagreeing in our highway conversation about memories of what the fence looked like. She referred to it as a picket fence and I said it was a tall board fence. To our amusement, after arriving in Hannibal this time, we found the reconstructed fence next to the famous house in the heart of that town's historic district was half and half, some of it representing each kind!

[107] A "perfect game" is a greater accomplishment than a "no-hitter" (in which one or more opposing batters may have walked, or reached base on an error, or by being hit by a pitch, but with no batter being credited with a base hit). Likewise, a "no-

hitter" is more perfect than a "shut-out," in which no opposing base runner scores, however many may have been left on base at the ends of one or more of the nine innings.

108 Mathematically, though almost impossibly improbable, a perfect game could involve as few as 27 pitches, if every batter swung at the first pitch but only hit it in such a way that he was put out. Or it could theoretically involve many more than 81 pitches if some batters hit a number of fouls, and some "took" up to three balls before striking out or otherwise being put out.

109 See, on the web, the Wikipedia article, "Prefect Game." Also, see "Perfect Games," on Infoplease.com.

110 Stories of this anomalous perfect game by two different sports writers appeared in the March 15, 2000 *Boston Herald*. Other stories appeared on that date's editions of *The Boston Globe, The New York Times, The Toronto Star*, and *The Toronto Sun*. Incidence of so-called "perfect games" in baseball has reflected social change in the larger society in which baseball is embedded. In the decades following World War II, for example, "the percentage of American women entering the labor force rose steadily" (see Henderson, 2006),and during the Second World War women in considerable numbers entered professional baseball—albeit not in the men's baseball leagues. The All-American Girls Professional Baseball League was formed in 1943 (under impetus from Chicago Cubs owner Philip Wrigley, of the chewing-gum company). As reported in the Ken Burns film about baseball, in all the seasons through which the league lasted two perfect games were recorded, both by one pitcher, Jean Faut of the South Bend Blue Sox. That league went out of existence after a few years, but female participation in the sport continues and there may in future be acceptance of women into major league ball clubs—just as the rosters have become increasingly interracial and international. One small indicator of that possibility: a perfect game was recorded by an 11-year-old girl, Katie Brownell, pitching for a boys Little League team in upstate New York on May 14, 2005. The Jersey she wore for that game was donated to the Baseball Hall of Fame, to be displayed either in its Youth Baseball exhibit or its Women in Baseball exhibit. For a list of major league perfect games see the MLB History website's "Rare & Memorable Feats."

111 As baseball has evolved, pitching tasks have become subdivided, so "pitchers" no longer is a unitary job category. There are somewhat different role expectations for "starting pitchers," "relief pitchers," and "closers." The latter two categories, collectively "the bullpen," are generally expected to work only one or a few innings per game, and thus may be called into two or more consecutive games, as the situation may necessitate, whereas "starters" are expected to work as many innings in a game as they can remain effective, thus needing more rest for their pitching arm between games, so they are part of a four- or five- man rotation, pitching only every fourth or fifth game. But in a preseason game such as that one between Boston and

Toronto the usual basis for a manager to change pitchers would not have applied and the use of six pitchers had little if any relation to this further specialization.

112 Although conscription for military service applied to men only in the United States through all wars so far, most people just prior to the middle of the 20th century still accepted that instance of gender discrimination as fair. Women were recruited for service in various non-combat roles in the U.S. armed forces in both world wars. The traditional division of labor by sex was bent but not truly broken in American forces in World War II. Combat roles for men only, women for non-combat roles only (except for nurses who could serve near areas of combat). Only recently have Americans come to regard it as fair to send both men and women "into harm's way." In some other countries it happened sooner.

113 In the western state small town where my formal education began, all my teachers but two were addressed as "Miss _____." I had a "Mr. _____" teacher in the woodshop for "Manual Training." The one teacher who was "Mrs. _____" was a young widow. My parents explained to me at the time that because her husband had died she was exempt from the then-legal exclusion of married women from teaching positions—a reflection, in part, of the effort in hard times to reserve too-scarce jobs for "heads of families" (i.e., men).

114 For a summary of what is known about the range of variation among human societies in their delineation of "men's work" and "women's work," see Van den Berghe (1973): Chapter 4.

115 Frequent news items about women in military service are indicative of this ongoing and sometimes problematic integration of the sexes. See, for example: Associated Press, January 2, 2004; Tucker, July 3, 2005; Parker, June 29, 2005. Also see such *New York Times* editorials as: May 24, 2003: "The Pinking of the Armed Forces"; May 20, 2005: "Chauvinism at the Battlefront" and May 29, 2005: "Disrespecting Women Soldiers."

116 See Wekesser and Polesetsky (1991):11-12.

117 Ibid., pp. 12-13

118 Saywell, pp. 156-164.

119 See Marshall 1987:4-12.

120 Nierenberg (2002) cited important demographic facts indicating problems arising from human over-reproduction, including the fact that population growth was still rapid in least developed countries, globally "the largest generation of young people in humanhistory—1.7 billion people aged 10 to 24" were about to enter their reproductive years, while more than 350 million women worldwide lack any access to family planning services, and over half a million women each year die from complications during pregnancy and childbirth.

121 As evidence for this change in attitude about reproduction the continuing decline in fertility rates later reached levels where various countries began to experience "negative population growth." As governments became concerned that declining

population could lead to diminution of national strength, some began trying to counteract the trend by instituting new incentives to reproduce. See Oleksyn 2006. As the transformation of gender roles has occurred, birth control technology has continued to advance and this has fostered increasing dissociation of sexual activity from reproduction. It is perhaps astonishing that sociologists, aware of the redefining of sex in recreational rather than procreational terms, have not yet seemed to link to these changes the "coming out" of homosexual persons and the change in public attitudes about homosexuality.

122 There is yet another question that will not be easy to answer by future research: Has that sex-role dedifferentiation in the occupational realm been due to (or incidental to) the increasing occupational specialization of the working population (quite apart from gender),and what part may it be playing in balancing or unbalancing the forces for societal integration versus societal disintegration? An answer to that question is not required for development of the thesis of the present book

Chapter 6

123 The new forms of wrongdoing characteristic of complex industrial societies with intricate exchange relations among interdependent specialized subgroups are not always visible and obvious. See Ross, 1907: 40-41. They "are to be discerned by knitting the brows rather than by opening the eyes. It takes imagination to see that bogus medical diploma, lying advertisement, and fake testimonial are death-dealing instruments. It takes imagination to see that savings-bank wrecker, loan shark, and investment swindler, in taking livelihoods take lives. It takes imagination to see that the business of debauching voters, fixing juries, seducing lawmakers, and corrupting public servants is like sawing through the props of a crowded grandstand. Whether we like it or not, we are in the organic phase, and the thickening perils that beset our path can be beheld only by the mind's eye."

124 Ibid., pp. 105-106.

125 Throughout the 1880s, consumers and the managers of railroad service were approaching agreement, albeit from diverse perspectives, that the national transportation system was just too chaotic and needed external control. In February 1887 both houses of Congress passed by large majorities an Interstate Commerce Act. President Cleveland signed it into law. Railroads were forbidden by it to discriminate among shippers. It required them to publish schedules of fares. It outlawed rebates. Price-fixing and traffic allocating pools were forbidden. And a five-member Interstate Commerce Commission was created to determine 'just and reasonable' rates. See the treatment of this topic in Strouse1999:255-257. Also see Morison 1965:763-764.

[126] This mistreatment of Ross by the Stanford administration eventually led to a "Report on Academic Freedom and Tenure," the charter document of the American Association of University Presidents, co-written in 1915 by Arthur Lovejoy, a Stanford philosopher who had resigned over Ross's firing, and Edwin R.A. Seligman, a Columbia University economist. The report asserted that self-governing faculty members, not university trustees, must decide standards for research and teaching, a view that was to become the fundamental principle of academic freedom. See the March 23, 2005 annual Benjamin N. Cardozo Lecture before the Association of the Bar of the City of New York given by Columbia University President Lee C. Bollinger, at<www.columbia.edu/cu/president/communications%20files/cardozolecture.htm>.

[127] Ross, op.cit.: 107-108.

[128] ibid.: 110-111.

[129] ibid.: 4-6.

[130] Urbina, 2006.

[131] Among economic and political theorists, mutualism denotes the belief that when labor is sold, or when a product of labor is sold, it should fetch in exchange a product or service produced by the same quantity of labor. An exchange that brings anything less would be exploitative, a theft, or "usury." The political movement of "Mutualists," followers of Pierre-Joseph Proudhon, advocated currency denominated in labor hours (labor notes) as a device for ensuring fair exchange. Perhaps a more accurate name for the movement would have been "equitablism."

[132] Ross, op.cit.: 3.

[133] Pérez-Peña, 2005.

[134] Ross, op.cit.: 6-10.

[135] ibid.: 10-11

[136] "The prosperous evildoers that bask undisturbed in popular favor have been careful to shun—or seem to shun—the familiar types of wickedness." ibid.:47.

[137] ibid.: 12-13.

[138] ibid.: 29.

[139] ibid.: 35-37.

[140] Johnson, 2006.

[141] McNeal, 2005.

[142] Homer-Dixon, 2005

[143] Lee, 2006

[144] "The upright may fall slack in devout observances, but he cannot afford to neglect his church connection. He needs it in his business. Such simulation is easier because the godly are slow to drive out the openhanded sinner who eschews the conventional sins. . . .He takes care to meet all the conventional tests,—flag worship, old-soldier sentiment, observance of all the national holidays, perfervid patriotism,

party regularity and support. Full well he knows that the giving of a fountain or a park, the establishing of a college chair on the Neolithic drama or the elegiac poetry of the Chaldæans, will more than outweigh the dodging of taxes, the grabbing of streets, and the corrupting of city councils." (Ross,1907:59-62)

[145] See Ogburn, 1966 (1922). As often happens, an incidental aspect of Ogburn's idea was too widely misconstrued as its essence. He illustrated the concept with an example of technological change occurring first, with folkways adapting to it only later, and many subsequent writers supposed cultural lag always meant advances in material culture(technology) always precede adaptive changes in nonmaterial culture. This neglects Ogburn's own statement (p. 211) that "forms of adaptation might be worked out prior to a change in the material situation." He was explicitly not implying that prior changes in nonmaterial culture would never occur which would necessitate subsequent adjustment of material culture. See also his later paper, "Cultural Lag as Theory" (reprinted in Duncan,1964).

[146] Ross, op.cit.: 69.

[147] ibid., pp. 84-85.

[148] ibid., pp. 97-98.

[149] See Lovelock, 2006, p. 137.

Chapter 7

[150] These two sentences start the second paragraph of the opening chapter of a textbook on *Social Disorganization* by Robert E. L. Faris (New York: Ronald Press Co., 1955, p. 3). At this departmental "Monster Rally," quoted as "inspirational," without context, the words evoked hearty and prolonged laughter. In context, however, they are not ludicrous. The paragraph preceding them ends with the sentence: "In order for us all to live and to get the various satisfactions of life, the relations of persons to one another have to have definite and enduring patterns of complementary relations—that is, organization." And the serious point being established at the beginning of this textbook chapter is further amplified by subsequent sentences: "[A human] requires a supporting organization of the type which is called a family to feed and care for him in infancy and to teach him the basic knowledge of household physics and mechanics . . . and to give him languages and the basic social sentiments and customs. He requires a school system to educate him, a police and law enforcement system to protect him, an immensely complex economic organization to supply material goods of life, a governmental structure to coordinate these, and a variety of organizations for other specific purposes."

Interestingly, the same fish-vs-human comparison that evoked laughter among those partying sociologists, was used in all seriousness some years later by an eminent historian, Barbara Tuchman. "All salmon," she wrote, "swim back to spawn in the headwaters of their birth; that is universal for salmon. But man lives in a more complicated world than a fish." See her *Practicing History: Selected Essays*. New York: Alfred A. Knopf, 1981.

The philosopher Daniel Dennett, to explain in Darwinian terms the evolution of freedom, also contrasted the instinctive behavior of this fish species with the cultural openness of human actions: "The salmon swimming upstream to spawn may be wily in a hundred ways, but she cannot even contemplate the prospect of abandoning her reproductive project and deciding instead to live out her days studying coastal geography or trying to learn Portuguese." See his *Freedom Evolves*, New York: Viking, 2003, p. 179.

[151] Even thousands of years ago, at various places around the world, groups of interacting human beings with specialized skills and some division of labor had devised more elaborate sailing craft: see Casson 1994. Remarkable feats of beyond-the-horizon sailing in small "primitive" vessels were achieved long ago. Moreover, in more recent times there have occurred impressive feats of rowing specially designed boats across both the Atlantic and Pacific oceans. See, for example, historical information available through the Ocean Fours Rowing Race website.

[152] Standard parlance in the U.S. Navy would call all vessels of such size and seaworthiness "ships," reserving the word "boat" to designate auxiliary craft small enough to be carried aboard a ship. Yet, Navy language had its own odd inconsistency, with submarines, as long as a football field, designated boats rather than ships.

[153] See, for example, Rosecrance, 1986; and Reich, 1991—especially Part Two: The Global Web.

[154] See McGrail 2002:311-345; Dodd 1972.

[155] See Leakey and Lewin 1992; and Stringer and McKie 1996.

[156] Darwin, 1937 [1845]:434-503.

[157] Even within the northern hemisphere, Britain's distance from North America was considerable in the days of sail. In 1620 it took the famous Mayflower sixty-six days to bring those Pilgrims across the Atlantic to settle Plymouth, Massachusetts. In 1809 the first steam-powered vessel to cross the Atlantic from America to Britain, the Savannah, took just under 30 days. (See Maxtone-Graham, 1972:3-4.)

[158] King 2003:237.

[159] Condliffe and Airey, 1953:178-181.

[160] King 2003:238.

[161] See Gibbs and Martin 1958, and 1962.

[162] Gibbs and Martin 1962.

[163] For impressions of the extensive division of labor just within the ship-building industry, see Maxtone-Graham 1972, Ch. 2, "Building the Mauretania." For vivid indications of division of labor within a ship's crew, and class distinctions among the passengers, see his Ch. 6, "Stokers & Steerage."

[164] Edmunds, 2000:64.

[165] Wilford 2003; Howard, 1987; Collins, 2003.

166 Skills of the building trades were similarly twice involved in a vastly different enterprise, the seven decades-long gestation of the famous Oxford English Dictionary, dedicated to tracing the evolving meanings of all words in the entire English language. See Winchester 1998:112-113, 149-150. Long-time editor of the project, James Murray, philological society president, had a corrugated iron shed (to be called the Scriptorium) built at the school where he taught, to house the accumulation, sorting, and organizing the dictionary's raw material—thousands and thousands of quotation slips sent in by volunteer readers of many, many books. Several years later when Murray left his position as a school teacher he had another, larger Scriptorium built in Oxford.

167 See King (2003):86-90.

Chapter 8

168 Emerson (1962:32)

169 Ibid., p.:33.

170 This is a milder but more broadly applicable statement than Ross's contention about being "at one another's mercy."

171 In conversation at the University of Wyoming, anthropologist William Mulloy told me in 1973 that humanity had been, in his view, "doomed since the Neolithic," i.e. Mulloy believed the Horticultural Revolution had put humankind on track toward the ultimate disaster of overshooting global carrying capacity. (This is a judgment shared and later eloquently portrayed by novelist Daniel Quinn.) From his own archaeological research on Easter Island Mulloy (1974:29) had concluded that its Polynesian inhabitants in pre-contact times had been "technologically successful" and "must have rejoiced in the solid assurance that their success was permanent" even though they were, in fact, facing catastrophic collapse from the damage their success had cumulatively done to their finite environment.

172 See *Sin and Society*, pp. 25-26

173 Ibid., pp. 35-37.

174 Adams (1993).

175 Portes and Sensenbrenner (1993).

176 Jones (1994a), (1994b).

177 Olzak et al. (1994).

178 In one of his earliest papers, Robert Merton (1934) similarly expressed concern that "unilinear evolution" in the direction posited by Durkheim's view of division of labor could continue until it would become self-defeating, with ever more frequent occurrences of anomie rather than any further increase of organic solidarity.

179 Vividly described in Blumberg's *The Predatory Society* (1989).

180 Frank (1995)

181 See Collier (1991); Schlesinger (1992); Cannon (1994); Raasch (1995).

[182] Light could be shed on that question, I believe, by considering anew Durkheim's reliance on Darwin for the rationale for his expectations concerning division of labor. Comparing human societies with natural ecosystems could be enlightening. In biotic communities, how does species diversity affect stability? What happens when predator-prey relations in a biotic community are altered by addition of new species populations?

[183] Stanley (1978):200.

[184] Schnaiberg (1980):227-231.

[185] See Iacocca (1984), especially pp. 34-38.

[186] Ibid., p. 10.

[187] Transcript, MacNeil/Lehrer News Hour on PBS telvision (Feb. 26, 1985:11-12).

[188] McCrone (1991).

[189] Smith (1983):23.

[190] Ross (1907):118.

[191] This episode occurred before cigarette ads were banned from television.

[192] See Desmond (1997):278-279.

Chapter 9

[193] Physicist Robert Adair, at Yale University, found that the sound may be either a "crack" or a "clunk," depending on whether the "sweet spot" of the bat hit the ball, and the difference tells the outfielder more immediately than he could infer from his initial reading of the ball's trajectory whether to run farther out or farther in. See Scientific American.com, June 11, 2001.

[194] Without having to think consciously of the geometry, outfielders run in a direction that has the visual effect of converting to a linear optical trajectory (LOT) the essentially parabolic path of the ball's actual travel. See McBeath, et al. 1995. See also, Gillies and Dodgson. 1999.

[195] Peterson 2002.

[196] There are many varieties of these aerial foragers: e.g., Alder flycatcher, ash-throated flycatcher, Hammond's Flycatcher, olive-sided flycatcher, Pacific sloped flycatcher, satin flycatcher, spotted flycatcher, Traill's Flycatcher. Barn swallows and mountain bluebirds also forage in the air. See Holmes and Recher (1986); Organic Farming Research Foundation (1999); Stein (1958). Some bats, too, are aerial foragers, using echolocation to home in on airborne prey. See, for example, Schnitzler, Moss, and Denzinger (2003).

[197] Dennett 2003: 171, elaborates upon the life-enabling predictabilities of our universe: "Mother Nature is not a 'gene-centrist.' That is, the process of natural selection doesn't favor transmitting information via genes when the same information (roughly) can be just as reliably, and more cheaply, provided by some other regularity in the world. There are the regularities supplied by the laws of

physics (gravity, etc.) and by the long-term stabilities of environment that can be safely 'expected' to persevere (salinity of the ocean, composition of the atmosphere, colors of things that can be used as triggers . . .). Since these conditions are more or less constant, they can be tacitly presupposed by the genetic recipes and not 'mentioned.'"

[198] For some insight into the various typewriter mechanisms and an appreciation of the technological advances they represent, see Russo 2002. The book includes a vivid chapter on connections between the typewriter's invention (in stages), emergence of a manufacturing industry, and proliferation of employment opportunities for women.

[199] See his *A Connecticut Yankee in King Arthur's Court*, a consistently imaginative and delightful yarn.

[200] Would we live more comfortably in a world where our incantations were just as effective as purposive and appropriately skilled actions? How would we prevent other denizens of such a world, also possessing such magical powers, from either deliberately or inadvertently misdirecting the processes we sought to direct?

[201] Catton 1980: 148

[202] Brooks 2008

[203] A partial exception to reliance on finger movements to activate the typewriter mechanism is the "automatic" typing machine, programmed to type copy after copy of a pretyped message. By individually typing in just the recipient's name and address, it could provide the illusion that a mass mailing was really a personal letter. Most of the finger movements in such mass typings simply occurred "further upstream."

[204] See Roberts 2004, p. 61 where Joseph Romm, who was assistant secretary for energy efficiency and renewable energy in the Department of Energy during the Clinton administration, is quoted as saying "if the U.S. government even brought up the possibility that global oil production might peak in, say, 2020, not only would that have an enormous and very negative impact on the markets, but it would essentially force the United States abruptly to change its energy policy to one that emphasized energy efficiency and alternative energy." The next administration after Clinton's was adamantly opposed to that kind of policy change.

[205] Catton 1980: 155. To clarify and illuminate the most fundamental dilemma confronting humanity in the 21st century it may be helpful to consider *Homo colossus* more than just metaphorically—as a distinct subspecies of humans, significantly different from pre-colossal *Homo sapiens*. That is something I will undertake in the last two chapters of this book.

[206] Mills 1958: 10-14

[207] That was not what agitated Mills. His concern was our apparent drift toward a (nuclear) Third World War. But see, for example, Brown, 2004; and Heinberg, 2003.

208 Uncertainty has been an important component of human experience and coping with it is a continuing element in human thought. In a chapter on "Moral Implications of Serendipity," Robert Merton and Elinor Barber (2004) said, "It is man's fate to recognize good and evil when it comes his way, and it is his lot as a valuing creature to try to establish the extent of his responsibility for it. The problem of unexpected good and evil is a part of the more general problem—the factor of unexpectedness has especial poignancy for men insofar as they place value on a predictable universe inhabited by rational individuals. Premature death and catastrophic illness, natural disasters, the frustration of efforts by 'bad luck,' all of these are to some extent incompatible with man's values, and to be borne they have to be explained as far as possible and justified. Good fortune, like ill fortune, is in some measure uncontrollable, and both create the problem of squaring things as they are with things as they should be, or, to put it another way, of answering the question of why they are as they are."

209 For a fascinating "biography of a scientific idea" (the greenhouse effect, which began to be recognized nearly two centuries ago) see Christianson 1999. To appreciate the recent penetration of that idea into public awareness, see Gore 2006. Also see news accounts of the international conference in Bali in December 2007.

210 A recent denial of the "climate change problem" was expressed by Michael Griffin, head of the National Aeronautics and Space Administration (NASA), as reported in *The New York Times*, June 1, 2007. Responding to a paper in the Journal of Atmospheric Chemistry and Physics by NASA and Columbia University scientists who contend that greenhouse gases produced by human activities have "brought the Earth's climate close to critical tipping points," Griffin took the position that it was "arrogant" for people of the present to suppose that the climate we have known "is the best climate for all other human beings."

Chapter 10

211 Sumner 1896; Lenski and Lenski 1982; Ostheimer and Ritt 1982.

212 Mountfort 1981; Singh 1984.

213 Sorokin 1957: 48.

214 Malinowski 1947: 27.

215 Parsons 1971: 142.

216 Singh 1984: 151.

217 ibid.: 140-141.

218 Schnaiberg 1980: 23-35.

219 For this global figure I have averaged energy-use by people in developed and underdeveloped countries combined. The estimate for per capita energy use just as the industrial era was about to begin was calculated by assuming half the population

of the world was then comparable to today's "South Asia Developing Countries" and half were comparable to all "Developing Countries," using food intake plus coal-equivalent energy consumption figures given in World Bank (1980: 420-424). The estimate for "today" was calculated from the figure for world total commercial energy consumption given in World Resources Institute and The International Institute for Environment and Development (1986: 104).

[220] Cottrell 1955: 1-14; Schnaiberg 1980: 21.

[221] Population density increased somewhat less because of U.S. territorial expansion in the meantime, but never mind, for comparison of population densities as such indicates anyway only part of what has happened to the ratio of load to carrying capacity. Presumably territorial expansion meant some increase of carrying capacity.

[222] Derthick and Quirk 1985.

[223] Leonard 1984.

[224] For formula used to calculate "cetacean equivalents" of human energy-use levels, see Catton 1986: 141

[225] Lumsden and Wilson (1983: 16

[226] This statement is, of course, a special case of the "tragedy of the commons." See Hardin 1968.

[227] DeBardeleben 1985: 15.

[228] Simpson et al. 1967: 101.

[229] See Allan (1965: 89) on "critical population density." Also see various references cited in Catton (1983: 278-284)—which traces development of the carrying capacity concept.

[230] (see Catton 1980: 214-217; or 1983: 272-275).

[231] For elaboration of this idea, see the section headed "From Exuberance to Post-Exuberance" in Catton and Dunlap (1980: 27-31).

Chapter 11

[232] Wittfogel (1968)

[233] Harris (1977: 155-163)

[234] See, e.g., Lovins and Lovins, 1982.

[235] See Goodell 2006

[236] Huang 1982; Poston and Yu 1985; Willis 1985a, 1985b.

[237] Parsons 1971: 57.

[238] Churchill and Lowe 1983.

[239] Arad and Arad 1979: 102-103.

[240] Lumsden and Wilson 1983: 33.

[241] Vayda 1974.

242 Heinberg 2006, pp. 54-55 says: "The Japanese attack on Pearl Harbor was triggered, at least in part, by the United States' decision to cut off oil exports to Japan in 1941, an action taken in response to Japanese pursuit of a Pacific empire. Japan, which had been almost completely reliant on imported oil, mainly from the US, concluded that it would have to obtain its oil elsewhere. This was a factor in its invasion of the oil-rich Dutch East Indies. "Historians of the Second World War now generally agree that Adolf Hitler planned to capture the oilfields of Romania by 1939 so that Germany would have a supply of oil. The next stages of the strategy included capture the oilfields of Persia by 1941, and those of Russia in 1942. However, America entered the war before these goals could be accomplished, and the allies managed to deny Hitler access to precious petroleum."

243 Garrett Hardin, interviewed by Hayes 1981: 66.

244 Daly 1979: 38.

245 See Ozinga 1985: 34. Oil is, of course, a "non-renewable" resource—meaning that oil deposits are geologically created from biomass at an enormously slower pace than the rate at which industrial societies extract them for use. Thus the very idea that people can be free to practice an oil-based way of life as long as the rate of discovery is not less than the rate of use is itself a form of ecological myopia. And yet, the idea that oil consumption rates need only be limited by discovery rates rather than by the rate at which the stuff is deposited in the earth can be implicit even in warnings by educated people that we must change our ways; a professor of petroleum engineering, for example, in a letter to a widely circulated newspaper (Marsden 1986), wrote: "In the period 1955-64, the world added three times as much oil to reserves as was consumed; from 1965 to '74, world reserves increased by twice as much as total consumption; but from 1975 to '84, when the incentive to find oil was unprecedented because prices were at an all-time high, the world, for the first time, consumed twice as much oil as was added to total reserves!" A few scholars have begun to contest the notion that "fossil fuels" all are derived from prehistoric biomass (see, e.g., Gold and Soter 1980), and further exploration of evidence bearing upon this idea will be relevant to an ecologically sophisticated sociology.

246 Ozinga 1985: 50.

247 See Hollin and Barry 1979. See also Cockburn 2007, Moulson 2007.

248 See Bach 1979; and Cowen 1986.

249 Laurmann 1979.

250 See Shabecoff 1986. See also Kay 2007.

251 Oppenheimer 1986.

252 Ledec 1985: 184. However, opening of "the northwest passage" by global warming (see BBC News 2007) will result for some countries in a seasonal alternative to the Panama Canal for access from the Atlantic to the Pacific and vice versa.

253 Nofz 1983.

254 Reining 1979; and Thomson 1985.

255 Milward 1977: 280-281.

256 Lloyd 1943.

257 Milward 1977: 239.

258 Campbell 1971.

259 Today, in many countries, a "women's movement" struggles to increase freedom of female entry into formerly all-male or mostly-male occupations. In evaluating prospects for future continued success of this struggle, we should note the irony of the following fact: female entry into formerly male roles was in the past a coercive change resulting from unwanted pressures. In occupied Germany in 1946, for example, the British Military Government required the holding of a legal job to qualify for a food ration card, and because of the postwar shortage of able-bodied males, women were conscripted for jobs that had formerly been male (Balabkins 1964: 170, 211).

260 Balabkins 1964.

261 Daoudi and Dajani 1985: 13-25.

262 Henderson 1978.

263 A wise British professor, who now teaches ethics at Oxford after starting out as an economist at the University of Bristol, has declared climate change is a serious ethical issue because it pits benefits for today's fossil fuel users against deprivations that will hit future generations facing fuel shortages, but (as an ironic holdover from the traditional expectation of perpetual progress) he nevertheless supposes posterity will be richer than the present population. See Broome 2008.

264 For the remaining life of anyone now living there will be a global carrying capacity deficit. The freedom of social scientists to build meaningful models of human society with no regard for the influence of ecological factors such as load pressure is gone. The forces that will diminish humans' freedom to breed, freedom to exploit natural resources, and freedom to dispose of metabolic products somewhere within the same environment in which we have to live, deserve sociological scrutiny. Scholars who become fully attuned to the importance of carrying capacity and other ecological concepts, could make a comprehensive, systematic, comparative study of rationing programs that would really illuminate the future. Some sociologists, however, adhering to vision-narrowing traditions of their discipline, are likely to be victims of an egregious cultural lag by ignoring or misconstruing the enormity of this change. An equivalent cultural lag is all too prevalent, as well, among the general public. Only if we pay increasing attention to the environmental and biological influences that shape human life (along with the social, cultural, economic, ideological and political forces with which people have heretofore tended to be preoccupied), can future adaptations by human societies to load pressure become less baffling and more predictable.

Chapter 12

[265] When my wife and I attended a showing of the clever Disney animation film Ratatouille, about the rat who aspired to be a world-renowned chef, only we and one other couple sat through the end-of-show rolling of the very long list of credits. It was an enlightening experience, enabling us to recognize that there were literally hundreds of people, in scores of specialized occupations, involved in making that cartoon film.

[266] I recently read of an admirable young man who for several years had pitched for a major league baseball team and was "let go" by that team. He was "picked up" by another team which signed him to a five year contract with a salary of several million dollars per year. Human person though he is, occupationally he is a tool for throwing baseballs. His marketable skill consists largely of throwing them in such ways that they are difficult for batters to hit in the ways well-paid batters are meant to do. Some would contend he is going to be overpaid by his new team, "earning" for each inning pitched as much as some of the fans watching him may earn in a year.

[267] Other items customarily contained in the same billfold, and even the billfold itself, may be of little interest to the pickpocket, or may even be anathema since they would identify their true owner as someone else. On the other hand, they may today be of prime importance to an identity thief, a role recently increased numerically. For such thieves the capture of, say, a Social Security card may be more desired than some coins and bills.

[268] Money obtained by illicit means may be preferred over theft of other commodities partly because money has a "one size fits all" quality—the money taken is exchangeable not just for one particular thing but for any of a variety of things. One successful pickpocket event may thus serve in lieu of several shoplifting events, reducing the risk of the shortcut taker being caught.

[269] See Dickens (1939): 79-81.

[270] See Dunn (1997): 111-114. Slang words used by pickpockets also tend to dehumanize various aspects of their "profession," not just its victims. Pickpockets often work in pairs, or sometimes trios, with one member of the team somehow distracting the mark while the other member picks the targeted pocket. The person doing the distracting may be called the stall. The one who actually picks the pocket is the cannon. The pocket to be picked has been called the kick. Success may depend upon "cleaning the cannon" immediately after the wallet has been removed from the kick, by surreptitiously turning the stolen item over to a third collaborator who quickly steps out of the scene. Thus the actual pickpocket, if apprehended, has no incriminating physical evidence in his possession. If a female participates in the event, she may provide the distraction by coming to stand in front of the mark

so that the cannon coming up behind can arrange a sudden collision. The mark, between the two of them, is caught in a sandwich, lasting only a second or two, during which the theft occurs.

271 For an example of losing by winning, see Dao 2005, and Mohajer 2007.

272 His punishment (being chained to a mountain and attacked daily by an eagle coming to eat his liver, which always grew back to be attacked again each following day, because he was immortal) can be regarded as an allegorical preview of the miseries that will befall humanity from global climate change. Was it just generic fire Prometheus gave to humans, or was it fossil fuel fire?

273 See Reuters (2002); Barrionuevo (2006); Lozano (2006).

274 For this explanation and interpretation, see Samuelson (2006).

275 See Johnson (2005).

276 See Lindgren's (2002) analysis of the "Know-Nothing CEO." For an attempt to comprehend the CEO's self-conception and his attitude about acquiring money, see Norris (2002).

277 See the Associated Press article, "Big 3 Execs to Talk Trade With Bush, *Washington Post*, Nov. 10, 2006.

278 See Thomas, "Bush to Meet With U.S. Automakers." Also see Krisher (2006)

279 This cartoon along with others expressing much the same derogatory view of the Chrysler-federal government relationship, was reprinted in the autobiography of the then head of Chrysler, Lee Iacocca.

280 See Iacocca (1984) Chapters 17, 18, and 19.

281 Ibid., pp. 199-200

282 Ibid., p. 225. But see also Porter (2006).

283 For example, see Hamilton (2005), and Hsu (2005).

284 See Skelton (2006).

285 In the last year before I retired from university teaching, I was appalled to find that my university had accepted advertising to be printed in the class time schedule document it distributed to students to enable them to make course selections each semester. I wrote to the administration to object on the ground that these ads degraded the class time schedule into a commercial document, one more instance of the "ulteriorism" one encounters in many everyday stimuli. Like the journalist who speaks of the "news hole" when referring to the portion of the newspaper not assigned to advertising, which conveys the idea that reporting significant happenings is incidental to selling products, I felt that the ad-infested class time schedule would become just one more degrading influence upon its student users. Choosing courses and course sections would be defined as incidental; being tempted to purchase advertised products becomes the ulterior function of the document. Inclusion of ads redefined the course schedule document as partly a service to merchants rather than wholly a service to students and the academic community. Thus, by implication, it redefined students as customers, not just seekers of learning.

286 Koop (1991): 163ff.

287 See Pringle (1998): 68.

288 Kessler (2001): 388.

289 My italics. Since there are millions of addicted smokers, simply making nicotine unavailable could be disastrous, Kessler (p. 392) acknowledged. But the corporate structure that promotes it, with all its political clout, could conceivably be controlled or even dismantled. He suggested the tobacco companies be spun off from whatever corporate parents now combine them with making and selling other, non-addictive, lines of merchandise. Congress could establish a tightly regulated non-profit corporation and assign it a monopoly over manufacturing and sales of the tobacco products existing addicts would demand. There have been numerous precedents, public bodies that provide transportation, or credit for home buyers, farmers, and students, etc. The new entity would supply tobacco products to those who want them, but without any promotion or incentives for sales. Similarly flagrant predatory practices occur in the pharmaceutical industry (see Petersen 2008), and in the modern world of finance (see Stein 1992).

Chapter 13

290 Just as truly, however, there is a need to avoid the opposite error. It is folly to suppose that all would be well for mankind if we could just learn to say the right words to each other. That misconception, too, has obscured for some of our contemporaries the situation actually facing mankind today. It is folly to suppose that anything like the "Consciousness III" advocated by Charles Reich (1970), or some other purely mental reorganization will suffice to achieve revolutionary improvement of man's lot in this world. The task before us must include a new focus on the power of language to shape perception and behavior, but it must avoid an obsessive supposition that linguistic factors are the exclusive determinants of human experience.

291 A cultural heritage is part of what Korzybski (1950) called "time binding."

292 And perhaps renewed and exacerbated by U.S. involvement in Iraq three decades after Vietnam.

293 As the scientific explanation of combustion advanced, our ancestors had to wean themselves from using the word "phlogiston," which had designated a supposed substance "given off by" whatever was being burned. When they learned to understand fire as a process of oxidation, they had to grasp the fact that burning fuels took on (and combined with) another substance, oxygen. The referent for the word "phlogiston" had been imaginary, but the word "oxygen" denoted something real. The vocabulary change was not just an arbitrary word replacement.

294 Borgstrom (1969):xi.

295 See Borgstrom (1969); Brown and Finsterbusch (1972); Hardin (1972); and Whelpton (1939).

[296] Vogt (1948): 44.

[297] A simple indication of this clinging to obsolete word-maps appeared on a reader board sign in front of the office of a construction company not far from my home. Someone in that company occasionally uses the movable letters to set up a 12-to-15 word homily to be read by the passing traffic. Recently it said: "The world has enough for every man's needs. But not enough for man's greed." Even the first sentence is today of doubtful ecological validity. The second sentence could only be truly meaningful if there were widespread agreement on the meaning of "greed." Few people ever see themselves as greedy. It is "the other guy, with more than we have" whom we see as greedy. But how much of the abundance "enjoyed" by *Homo colossus* is really necessary to the life of *Homo sapiens*?

[298] See Catton 1980: 51-52.

[299] See the chapter by Lloyd V. Berkner and Lauriston C. Marshall in Brancazio and Cameron 1964. For a less technical version, see the article by the same authors in the tenth anniversary issue of *Saturday Review*, May 7, 1966.

[300] The implications of this basic fact were abundantly clarified in a too little-known book by Cottrell (1955).

[301] This refers to an average rate of storage over the past billion years. Earlier on, carbon was probably sequestered by natural processes at a considerably faster rate, and was being sequestered more slowly in, say, the most recent 50 million or 100 million years. But the point holds, that our rate of undoing nature's sequestration is fantastically rapid. Current proposals for what ought to be called re-sequestration seem too feeble to remedy the damage we do.

[302] It should be noted that I am here using the word production more literally (to mean the actual forming of coal and petroleum and natural gas) than the usual usage which calls extraction "production."

[303] For examples of pioneering efforts to update the world's word-maps about energy before it was too late, see Cottrell (1955), Hubbert (1969), and Odum (1971).

[304] That we have been living mainly on energy imported from antiquity is among the facts commonly resisted as if their "inconvenience" necessarily made them irrelevant, or even false. See Catton (1996),

[305] For the extent of oil depletion and implication thereof, see Hubbert (1969): 180; Heinberg (2003); Deffeyes (2005).

[306] He made the suggestion in response to critics of the handful of experimental blasts previously set off there. For a more recent stubbornly Cornucopian view of "endlessly available" fossil energy, see Huber and Mills 2005.

[307] See Moorcraft (1973).

[308] This was emphatically demonstrated by Sears (1935), and echoed by Borgstrom (1969).

[309] See Starke (2007).

[310] For an indication that this was foreseen before it seemed serious, see Oestreicher (1973). Now it has begun to dawn on some thoughtful writers that catastrophic weather events will be increasingly common in the reshaped meteorological future wrought by *Homo colossus*. See, for example, the Jan. 9, 2009 editorial "Is this the future of Northwest winters?"

Chapter 14

[311] Vogt (1948):146.
[312] See Schnaiberg (1980).
[313] Quoted by Dorst (1970):15.
[314] Lynd & Lynd (1929):225.
[315] See Lomborg (2001), and Simon and Kahn (1984).
[316] For example, see comments by E. O. Wilson, Stephen H. Schneider, Norman Myers, Lester R. Brown, Emily Matthews, Al Hammond, Devra Davis, David Nemtzow, and Kathryn Schultz in *Grist* 2001.
[317] Maddox (1972):78.
[318] Lynd and Lynd (1929):498.
[319] Maddox (1972):7, 44, 108, 257, 274.

Chapter 15

[320] Ambrose (1998). Regarding the applicability of this biological concept to the human prospect, consider the thrust of Jared Diamond's writings. As stated on the page opposite its Table of Contents, the "theme" of his The Third Chimpanzee (1992) is "How the human species changed, within a short time, from just another species of big mammal to a world conqueror; *and how we acquired the capacity to reverse all that progress overnight*" (my emphasis). Two of Diamond's subsequent books amplify this. See his *Guns, Germs and Steel* (1997a), and his magisterial coverage of factors leading to reversal in *Collapse* (2005). A fourth book by Diamond, titled *Why Is Sex Fun?* (1997b) is no less truly concerned with evolutionary theory, albeit a side-branch off the ominous main theme represented by the other three books.
[321] Catton (1984).
[322] Anable (1975).
[323] Strother & Methvin (1975). For some more recent views about terrorism and the media, see Paletz & Schmid (1992), and Martin & Petro (2006).
[324] Mauss (1975).
[325] Wilson (1968):116.
[326] Aldridge (1969):93-94.

327 In his first State of the Union address, President Richard Nixon noted this
diminution of faith, and that allusion to it was discussed by Stewart Alsop in "The
Mysterious American Disease," in Newsweek, February 9, 1970. The decline was
exemplified especially in Heilbroner (1974).

328 See Webb (1952).

329 Potter (1954).

330 Catton (1976).

Chapter 16

331 Most notably, Walt W. Rostow (1963) devised a theory of "economic development"
in which this metaphor was a focal concept.

332 See Deffeyes 2005.

333 See Milbrath 1989; Ophuls and Boyan 1992. There is too much of this whistling in
the dark as we approach a graveyard. Had billions and billions of tons of formerly
atmospheric carbon not been extracted from the air by early life forms and buried
underground, where they were transformed by geological processes over hundreds
of millions of years, it would have been impossible for animal life (and human
life) to evolve on this planet. Now, when adequate account is taken of knowledge
acquired in biology and chemistry, it becomes evident that our not-so-long-ago
human ancestors made a tragic error when they first defined all that buried carbon
as fuel. In retrospect, we should see that it was nature's fortuitous sequestration of
a substance whose removal from the primordial atmosphere made us possible.

334 See Roberts 2004; Deffeyes 2005; Howe 2006.

335 See Wald 2006, and Associated Press 2007.

336 See Cox 2005

337 See Ehrlich and Ehrlich 1981: 129-176 and 199-206.

338 See Heinberg 2006:94.

339 See Wackernagel and Rees 1996.

340 The concept of a "biological species" is defined as a population that is reproductively
isolated from other populations, but when reproductive isolation cannot be
ascertained, or is not complete, descendants of a particular species population often
exhibit a markedly changed pattern of resource consumption, changed metabolism,
etc., so that they function quite differently as an ecosystem component. They may
then be regarded as a new species, or a new subspecies. As hominids evolved to
the level where cultural heritage began to be as important as genetic heritage in
shaping their lives, that change was highly significant and made it reasonable to
see them as a new species or subspecies, whether or not they were reproductively
isolated from other hominid contemporaries. My use of the term *Homo colossus* to
refer to people in the industrial societies of today is based on regarding tools and
elaborate technological apparatus as prosthetic extensions of human bodies. People

thus equipped by culture are thereby enabled to live by lavish exosomatic (outside the body) energy use and by the exosomatic consumption of both biotic and abiotic substances. Regarding them as a distinct subspecies enables us to recognize that different carrying capacity limits apply to them than to non-colossal members of the species *Homo sapiens*. The 21st century bottleneck will affect both *H. colossus* and the rest of *H. sapiens*, but we need to understand that it will affect these two distinguishable subspecies differently. On the varieties of species definitions, see Zimmer (2008).

341 See Simmel 1978. In societies with extensive division of labor, the products, services and policies money can buy are highly varied, and their "value" (price) depends on the relation between both present and expected future demand and supply. "Demand" means having a desire to purchase, together with possessing, or having access to, the "wherewithal." So effective demand depends on both present and anticipated future possession of money. Whatever a particular society's economic religion, acquiring money has become a major theme of modern life. Escalating division of labor inevitably led to that. To acquire money was at first to acquire a means of obtaining desired products, services and actions. But the pervasive pursuit of money eventually reached a higher level of abstraction in our lives. As a means of storing value, money-accumulation can become a goal sought with very little conscious and deliberate attention to specific intended expenditures. All that has been said in these chapters about the dehumanizing influence of division of labor has intentionally gone far beyond the concern expressed by Marx (and later social scientists following his notions) about worker alienation (see Lo 2000, Blauner 1964, Braverman 1974). It is my contention that the monetizing-of-people influence of the exchange relations necessitated by industrial-level division of labor tends to alienate everyone from everyone else—of course in varying degrees, and in most cases, perhaps, only somewhat. I believe this effect runs much deeper than was apparently felt by Lipset and Schneider 1983. Modern division of labor has made us less likely to share the belief of John Donne from long ago, that "Every man's death diminishes me." Consider, for example, the young soldier from Groveland, California, who died in the crash of his Black Hawk helicopter 180 miles north of Baghdad. It was newsworthy when his wife's mother in Michigan fought back against these modern currents of dehumanization. She described to the *Muskegon Chronicle* her view of the loss of her son-in-law, father of her two grandchildren. His death destroyed a family, she told the paper. "He wasn't just a number. I want people to know Matthew Tallman existed. He was a good father, good husband and good human being." See Gilbert 2007. To want to have it known that we existed—and that our existence mattered—is a very human urge, probably not felt, much less ever expressed, by our nonspeaking chimpanzee cousins.

342 Although the financial-system collapse that hit the world in 2008 and so deepened the economic recession that the incoming American administration in 2009 was

expected to give absolute first attention to "stimulating the economy" so as to bring about urgently desired "recovery," it remained doubtful that recovery would be swift—or even possible (see Heinberg 2009). Insistence that legislative action to "create several million jobs" could fortuitously put humanity's use of the biosphere on a "greener" basis was not altogether persuasive in hard times. Recession turning to depression, and perhaps non-recovery, may have to be understood in ecological terms—as the initial aspect of the 21st century bottleneck.

GLOSSARY

ALLOPATRIC: occurring or existing in separate places that are isolated from one another (as in "allopatric speciation," development of distinct daughter species when populations of an ancestral species have become geographically isolated from each other and evolve under different environmental conditions).

ANOMIE: prevalence of the view that antisocial types of behavior are necessary as means to attain socially prescribed goals.

ARCHAEOPTERYX: a genus of extinct winged, feathered creatures with teeth—intermediate between reptiles and birds.

AUSTRALOPITHECUS: a genus of extinct ape-like primates, considered ancestral to various hominid species, including (eventually) *Homo sapiens*.

AUTOTROPHS: organisms capable of photosynthesis (or, in some cases, chemosynthesis), which can convert abiotic substances into energy-rich organic substances.

BIOGEOCHEMICAL CYCLES: patterns of circulation of energy and a score of chemical elements involved in life processes, in systems comprising earth, water, air, and organisms. The chemical materials can be and are recycled over and over again; the energy cannot be recycled.

BIOMASS: the amount of organic material in some specified context.

BIOSPHERE: the region at and near the Earth's surface within which life occurs, including land, water, and air; all life on the entire planet.

BIOTIC COMMUNITY: a more or less self-sufficient and localized web of life collectively adapting to the life-supporting conditions of the local environment.

BOTTLENECK: a drastic narrowing of life opportunities, especially when an expanding population has overused its environment, bringing about a carrying capacity deficit, so that in subsequent generations the population decreases rather than continuing to increase.

CARNIVORE: an animal type that subsists by eating the flesh of other animals

CARRYING CAPACITY: the maximum population of a given species which a particular environment can support indefinitely (in the case of the human species, under a specified technology and organization; hence, the maximum sustainable ecological *load*—population X per capita resource use).

CARRYING CAPACITY DEFICIT: the condition wherein the permanent ability of a given environment to support a given form of life is less than the quantity of that form already in existence and living by use of that environment.

CARRYING CAPACITY SURPLUS: the condition wherein the permanent ability of a given environment to support a given form of life exceeds the quantity of that form then in existence.

CRASH: the more or less precipitate decline in numbers that follows when a population has exceeded the carrying capacity of the environment available to support it.

CULTURAL LAG: stress that occurs when some of the patterns of culturally prescribed behavior or belief change more than other interconnected patterns.

CULTURE: a system of socially acquired and transmitted standards of judgment, belief, and conduct, as well as the social and material products of the resulting conventional activities.

DEHUMANIZATION: implicit denial of or disregard for whatever traits are conventionally regarded as essential to being human; systemic reduction of people to the status of tools, instruments, resources, commodities.

DETERMINISM: a system of reliable and discoverable connections between events, between actions and results, between causes and consequences.

ECOSYSTEM: a comprehensive web of interrelations between organisms, other organisms, and their environment; it tends to be characterized by the operation of various checks and balances.

ENERGY SLAVES: a metaphor by which the value of fossil energy is measured in terms of the human muscle-power equivalent to it. (For example, burning about 32 U.S. gallons (121 liters) of gasoline releases energy equal to the energy content of the food an active human adult would consume in a year. Assuming comparable energy-conversion efficiency by fuel-burning machinery and food burning human bodies, quantitatively the work done by burning that much gasoline in an engine tends to approximate a year's work by a human slave.)

ETHNOCENTRISM: the tendency to regard the behavior of people raised by other standards than our own as wrong, rather than just different.

EVOLUTION (biological): imperfect replication of the traits of organisms in their progeny and the selective retention among descendant populations of those traits best adapted to prevailing environmental circumstances.

EXOSOMATIC METABOLISM: external energy-conversion processes (e.g., in mechanical apparatus involved in modern ways of living, especially by *Homo colossus*) that supplement the physical and chemical processes of life that occur within human bodies.

FATE: in human experience, whatever happens to us that is uninfluenced by our own actions. (A special case, Anthropogenic "Fate," occurs when circumstances not intended by anyone occur as the result of innumerable small decisions about other matters by innumerable people.)

FOSSIL FUELS: energy-rich combustible substances created by geological transformation of the remains of organisms that lived long ago, including coal, petroleum, and natural gas.

GHOST ACREAGE: amount of additional farmland that would be required to yield biofuels equivalent in energy content to what is being obtained from fossil fuels.

HERBIVORE: an animal type that subsists by eating plant tissues.

HETEROTROPH: an organism which cannot convert abiotic substances into organic matter but must use for its sustenance materials produced by other organisms.

HOMINID: a member of the family *Hominidae*; a human being, or a member of the various extinct species descended from ape-like ancestors and having some human-like traits.

HOMO COLOSSUS: modern human beings equipped with technology (tools or apparatus) that greatly enlarges their per capita resource demands and environmental impact.

HOMO NEANDERTHALENSIS: the most recently extinct human species, once regarded as a subspecies of *Homo sapiens* but now not so considered, nor ancestral to modern *Homo sapiens*. The two species did co-exist for a time in southeastern Europe and especially in the Middle East.

HOMO SAPIENS: the species of mammal that includes both the reader and the author, and all other contemporary human beings; a language-using, tool making, social species, descended from earlier types of humans who were also members of the genus *Homo*, capable of evolving culturally as well as genetically.

HUBRIS: excessive pride; excessive self-confidence; excessive sense of one's own competence.

MAGIC: supposed achievement of results be means that are in fact inadequate to produce such outcomes; opposite of determinism.

MECHANICAL SOLIDARITY: in sociology, the type of social cohesion attributable to shared norms and values among all members of a society. Generally seen as characteristic of small, pre-industrial societies.

METABOLISM: in general, chemical transformations of ingested substances making the life of an organism possible (including both *anabolism*, the process by which food is turned into living tissue, and *catabolism*, the process by which energy is released to support the organism's activity while tissue is turned into waste products). See also EXOSOMATIC METABOLISM.

MUTUALISM: a strong and reciprocal interdependence between different but associated life forms.

NATURAL SELECTION: the favoring of one variant over another by its more advantageous adaptation to existing environmental circumstances, as measured by differential rates of replacement in succeeding generations. (See selection pressure.)

NICHE: the role that an organism (of a given kind) plays in an ecosystem—i.e., the kinds of resources it has to obtain from its environment, the kinds of things it must do to its environment in the process of living, and the kinds of relationships it must have with other organisms to go on living.

ORGANIC SOLIDARITY: in sociology, the kind of social cohesion that has been expected to arise from interdependence among occupationally differentiated members of a society. Generally supposed to characterize modern industrial societies with extensive division of labor.

OVERPOPULATION: population in excess of its environment's carrying capacity for its way of living.

OVERSHOOT: (v.) to increase in numbers so much that the available carrying capacity is exceeded by the ecological load, which must in time decrease accordingly; (n.) the condition of having exceeded for the time being the permanent carrying capacity of an environment.

PHOTOSYNTHESIS: a process (occurring mainly in plants and certain microorganisms) by which the energy of sunlight, enabled by chlorophyll, synthesizes organic molecules from carbon dioxide and water, and releases oxygen.

PRIMATES: an order of placental mammals that includes lemurs, lorises, monkeys, apes, and humans.

QUASI-SPECIATION: the non-genetic differentiation of a human population into differently specialized subgroups by use of alternative tools, customs, or symbols.

REDUNDANCY-ANXIETY: a morbid apprehension fostered by carrying capacity deficit when each person is challenged by the latent worry that he or she may be part of the load surplus.

RENEWABLE RESOURCES: usable substances produced by on-going processes such as organic growth that occur at rates commensurate with actual or potential rates of consumption; e.g., usable energy obtained directly or indirectly from contemporary solar inputs, rather than withdrawn from finite quantities accumulated from past solar inputs.

SELECTION PRESSURE: any circumstance that causes one variant to be more successful than another in surviving and producing descendants. (See natural selection.)

SIGNIFICANCE DEPRIVATION: experienced lack of respect; a feeling that one has no impact on others or no influence upon events.

SIN: as defined for modern times by sociologist E. A. Ross—any human action that harms other humans, directly or indirectly.

SPECIATION: differentiation of a population into distinct types that do not interbreed and that have genetically-based traits adapted to different environmental conditions. (*Allopatric speciation* refers to geographically separated subpopulations of an ancestral species giving rise to several "daughter" species through adaptation to *different* environments. *Sympatric speciation* would involve various "daughter species" somehow diverging from a common ancestral species within the *same* environment.)

SPECIES: a category of organisms taxonomists have judged to be sufficiently distinct from others for recognition as a separate kind; assumed (or known) to be incapable of interbreeding with another species.

SUCCESSION: an orderly and directional process of change in the composition of a biotic community, resulting from effects of its life processes upon its environment. As former member species dwindle and die out they are replaced by other species (with access to that locality but not previously living there) which happen to be better suited to the changed conditions.

SYMPATRIC: occurring or existing within a single territory (as in hypothetical "sympatric speciation.")

TERRITORIALITY: an animal behavior pattern that commonly arises in response to actual or potential resource shortages: individuals or groups lay claim to distinct territories for feeding or breeding, and drive off competitors to ensure that scarce resources will adequately support the claimant population.

List of Sources

Adams, Julia.
> 1993. "Working-Class Politics in Nineteenth-Century Toulouse, France: Paths of Proletarianization Revisited." *Social Science History*, 17 (Summer):195-225.

Aldridge, John. W.
> 1969. "In the Country of the Young." Harper's Magazine (November).

Allan, William.
> 1965 *The African Husbandman*. Edinburgh: Oliver and Boyd.

Alsop, Stewart.
> 1970. "The Mysterious American Disease." Newsweek (February 9).

Ambrose, Stanley H.
> 1998. "Late Pleistocene human population bottlenecks, volcanic winter, and differentiation of modern humans." Journal of Human Evolution 34 (6): 623-651.

Anable, David.
> 1975. "Coming to Grips with World Terrorism." Christian Science Monitor, Dec. 19.

Anderson, Walter Truett.
> 1996. Evolution Isn't What It Used to Be: The Augmented Animal and the Whole Wired World. New York: W. H. Freeman and Co.

Angier, Natalie
> 2005. "Independently, Two Frogs Blaze the Same Venomous Path." The New York Times, August 9.

Anonymous
> n.d. "Economic and human costs of the Jyllands-Posten Muhammad cartoons controversy." From *Wikipedia, the free encyclopedia* (on the Internet).

Anthony, Jean.
> 1981. "Milne-Edwards, Henri." vol. 9, pp. 407-409 in Charles Coulston Gillispie (ed.), Dictionary of Scientific Biography. New York: Charles Scribner's Sons.

Arad, Ruth W., and Uzi B. Arad.
> 1979. "Scarce Natural Resources and Potential Conflict." pp. 23-104 in Ruth W. Arad, Uzi B. Arad, Rachel McCulloch, Jose Pinera, and Ann L. Hollick, Sharing Global Resources. New York: McGraw-Hill.

Ash, Roberta
> 1972. Social Movements in America. Chicago: Markham Publishing Company.

Associated Press.
> 2004. "Female Casualties in Iraq Not Played Up." The New York Times, Jan. 2.

> 2006. "Big 3 Execs to Talk Trade With Bush." Washington Post, Nov. 10.

> 2007 "Ky. Pilots Missed Warnings Before Crash." Washington Post, July 22.

Bach, Wilfrid
> 1979 "Impact of Increasing Atmospheric CO2 Concentrations on Global Climate: Potential Consequences and Corrective Measures." Environment International, 2: 215-228.

Baker, Al.
> 2005. "Crime Numbers Keep Dropping Across the City." The New York Times, Dec. 31.

Balabkins, Nicholas
    1964 Germany Under Direct Controls: Economic Aspects of Industrial
        Disarmament 1945-1948. New Brunswick, NJ: Rutgers
        University Press.

Barlow, Connie C.
    1997 Green Space, Green Time, New York: Copernicus

Barrionuevo, Alexei.
    2006. "Ex-Enron Chief Is Sentenced to 24 Years." The New York
        Times, October 23

BBC NEWS.
    2006. "Muhammad cartoon row intensifies." Feb. 1.

    2007 "Warming 'opens Northwest Passage'." Sept. 14.

Becker, Carl Lotus.
    1922. The Declaration of Independence: A Study in the History of
        Political Ideas. New York: Harcourt, Brace.

Bergman, Charles, and Kathryn Fontana
    2009. OpEd—"Oil's human cost: Ecuador trip reveals how energy
        demand is poisoning a country." Tacoma, WA: The News Tribune,
        January 11.

Blair, Frank.
    1959. "Ecology and Evolution." Antioch Review. 19 (Spring):47-55.

Blau, Peter M.
    1977. Inequality and Heterogeneity: A Primitive Theory of Social
        Structure. New York: The Free Press

Blauner, Robert.
    1964 Alienation and Freedom: The Factory Worker and His Industry.
        Chicago: University of Chicago Press.

Blumberg, Paul.
    1989. The Predatory Society: Deception in the American Marketplace.
        New York: Oxford University Press.

Bonner, John Tyler.
  1988. The Evolution of Complexity: By Means of Natural Selection.
      Princeton, NJ: Princeton University Press.

Bookchin, Murray
  1982 The Ecology of Freedom: The Emergence and Dissolution of
      Hierarchy. Palo Alto, CA: Cheshire Books.

Borgstrom, Georg.
  1959. Too Many: A Study of Earth's Biological Limitations. New York:
      Macmillan.

Bourne, Geoffrey H.
  1962. Division of Labor in Cells. New York and London: Academic
      Press.

Bradley, Harriet
  1989 Men's Work, Women's Work: A Sociological History of the Sexual
      Division of Labour in Employment. Minneapolis: University of
      Minnesota Press

Brancazio, Peter J., and A. G. W. Cameron (eds.)
  1964. The Origin and Evolution of Atmospheres and Oceans. New
      York: John Wiley and Sons.

Braverman, Harry.
  1974 Labor and Monopoly Capital. New York: Monthly Review Press.

Brockman, John (Ed.)
  2006 Intelligent Thought: Science versus the Intelligent Ddesign
      Movement. New York: Vintage Books

Brokaw, Tom
  1998 The Greatest Generation. New York: Random House.

Brooks, David
  2008 "The Behavioral Revolution." The New York Times, October 28.

Broome, John
  2008 "The Ethics of Climate Change." Scientific American, 298
      (June):96-102.

Brown, Lester R.
> 2004. Outgrowing the Earth: The Food Security Challenge in an Age of Falling Water Tables and Rising Temperatures. New York: W. W. Norton & Co.

Brown, Lester R., & Finsterbusch, Gail W.
> 1972. Man and His Environment: Food. New York: Harper & Row.

Browne, Janet.
> 1996. Charles Darwin: Voyaging. A Biography. Princeton, NJ: Princeton University Press.

> 2002. Charles Darwin: The Power of Place. Volume II of A Biography. New York: Alfred A. Knopf.

Campbell, Colin
> 1971 Wage-Price Controls in World War II, United States and Germany. Washington, DC: American Enterprise Institute.

Cannon, Angie.
> 1994. "Fed Up: A Less Kind, Less Gentle America Emerges in Poll of Nation's Mood." The News Tribune, Tacoma, WA. Sept. 21, p. A3.

Casson, Lionel
> 1994. Ships and Seafaring in Ancient Times. Austin: University of Texas Press.

Catton, William R., Jr.
> 1976 "Why the Future Isn't What It Used to Be (And How It Could Be Made Worse than It Has to Be)." Social Science Quarterly, 57 (September): 276-291

> 1980. Overshoot: The Ecological Basis of Revolutionary Change. Urbana: University of Illinois Press.

> 1983. "Social and Behavioral Aspects of the Carrying Capacity of Natural Environments." pp. 269-306 in Irwin Altman and Joachim F. Wohlwill (eds.), Behavior and the Natural Environment. New York: Plenum.

1984. "Probable Collective Responses to Ecological Scarcity: How Violent?" Sociological Perspectives, 27 (January): 3-20.

1986 "Homo colossus and the Technological Turn-Around." Sociological Spectrum, 6: 121-147.

1996. "The Problem of Denial." Human Ecology Review, 3 (Autumn):53-62.

Catton, William R., Jr., and Riley E. Dunlap
1980 "A New Ecological Paradigm for Post-Exuberant Sociology." American Behavioral Scientist, 24 (September/October): 15-47.

Childs, Marquis W., and Douglass Cater
1954 Ethics in a Business Society. New York: The New American Library of World Literature, Inc.

Christianson, Gale E.
1999 Greenhouse: The 200-year Story of Global Warming. New York: Walker and Company.

Churchill, R. R., and A. V. Lowe
1983 The Law of the Sea. Manchester, UK: Manchester University Press.

Cloward, Richard A., and Lloyd Ohlin.
1960 Delinquency and Opportunity. New York: The Free Press.

Cockburn, Alexander
2007 "Is Global Warming a Sin?" The Nation, May 14.

Codevilla, Angelo M.
1997 "Editor's Introduction." pp. vii-xviii in Angelo M. Codavilla (editor and translator), The Prince by Niccolò Machiavelli. New Haven: Yale University Press.

Cohen, I. Bernard
1980 The Newtonian Revolution: With Illustrations of the Transformation of Scientific Ideas. New York: Cambridge University Press

Collier, James Lincoln.
    1991. The Rise of Selfishness in America. New York: Oxford University
    Press.

Collins, Mary.
    2003. AIRBORNE: A Photobiography of Wilbur and Orville Wright.
    Washington, DC: National Geographic

Condliffe, John Bell, and W. T. G. Airey
    1957. A Short History of New Zealand. Christchurch: Whitcombe
    & Tombs.

Cooley, Charles Horton.
    1920. "Reflections Upon the Sociology of Herbert Spencer." American
    Journal of Sociology, 26(September):129-145.

Connolly, Ceci.
    2006. "U.S. Plan For Flu Pandemic Revealed." Washington Post, April
    16.

Cortazal, Manuel.
    2006. "Preparing for pandemic." Boston Globe, March 17.

Coser, Lewis A.
    1977 Masters of Sociological Thought (Second edition). New York:
    Harcourt Brace Jovanovich, Inc.

Cottrell, Fred
    1955 Energy and Society. New York: McGraw-Hill.

Cowen, Robert C.
    1986 "Global Warming Compels Scientists to Close Ranks." Christian
    Science Monitor, January 15, p. 1.
Cox, John D.
    2005 Climate Crash: Abrupt Climate Change and What It Means for
    Our Future. Washington, DC: Joseph Henry Press

Dalton, Patricia.
    2006. "The Don't Blame Me Generation." Washington Post, March
    5, p. B-3.

Daly, Herman E.
  1979 "Ethical Implications of Limits to Global Development." pp. 27-57 in William M. Finnin, Jr. and Gerald Alonzo Smith (eds.), The Morality of Scarcity: Limited Resources and Social Policy. Baton Rouge: Louisiana State University Press.

Dao, James
  2005. "Instant Millions Can't Halt Winners' Grim Slide." The New York Times, December 5.

Daoudi, M. S., and M. S. Dajani
  1985 Economic Diplomacy: Embargo Leverage and World Politics. Boulder, CO: Westview Press.

Darwin, Charles.
  1859. On the Origin of Species by Means of Natural Selection. London: Charles Murray. (Final edition: New York: Mentor Books, 1958 [1872].)

  1937 [1845] The Voyage of the Beagle [Harvard Classics edition]. New York: P. F. Collier & Son Corporation

DeBardeleben, Joan
  1985. The Environment and Marxism-Leninism: The Soviet and East German Experience. Boulder, CO: Westview Press.

Deffeyes, Kenneth S.
  2005. Beyond Oil: The View from Hubbert's Peak. New York: Hill and Wang

Dennett, Daniel
  2003. Freedom Evolves. New York: Viking

Derthick, Martha, and Paul J. Quirk
  1985. The Politics of Deregulation. Washington, DC: The Brookings Institution.

Desmond, Adrian.
  1997. Huxley: From Devil's Disciple to Evolution's High Priest. Reading, MA: Addison-Wesley

Desmond, Adrian, and James Moore. 1991. Darwin. New York: Warner Books, Inc.

Diamond, Jared.
    1992 The Third Chimpanzee: The Evolution and Future of the Human Animal. New York: HarperCollins

    1997a Guns, Germs, and Steel: The Fates of Human Societies. New York: W. W. Norton & Co.

    1997b Why Is Sex Fun?: The Evolution of Human Sexuality. New York: HarperCollins.

    2005. Collapse: How Societies Choose to Fail or Succeed. New York: Viking

Dickens, Charles
    1859 A Tale of Two Cities. London: All the Year Round

    1939. The Adventures of Oliver Twist. New York: The Heritage Press.

Dodd, Edward
    1972. Polynesian Seafaring. New York: Dodd, Mead.

Dopp, Katharine Elizabeth.
    1904. The early cave-men. Chicago : Rand, McNally & company

Dorst, Jean.
    1970. Before Nature Dies (transl. by Constance D. Sherman). Boston: Houghton Mifflin.

Duncan, Otis Dudley (ed.).
    1964. William F. Ogburn on Culture and Social Change. Chicago: University of Chicago Press.

Dunn, Jerry.
    1997. Idiom Savant: Slang As It Is Slung. New York: Henry Holt and Company.

Durkheim, Emile.
    1984. The Division of Labor in Society. (Transl. by W. D. Halls). New York: The Free Press. [original French edition, 1893]

Edgren, Hobart
    1959 Of Marble and Mud. New York: Exposition Press

Editorial
    2009 "Is this the future of Northwest winters?" Tacoma, WA: The News Tribune, January 9.

Edmunds, Peter.
    2000. Vivat Heathrow! Airport Life Unwrapped. Kington Magna, Gillingham, Dorset, UK: Cirrus Associates

Eddy, Melissa.
    2006. "Bird Flu Found in Cat in Germany." Washington Post, Feb. 28.

Efron, Sonni.
    2006. "Google-Earthing the Hermit Kingdom." Los Angeles Times, 29 Aug.

Ehrlich, Paul R.
    1986 The Machinery of Nature. New York: Simon and Schuster

Ehrlich, Paul R., and Anne Ehrlich
    1981 Extinction: The Causes and Consequences of the Disappearance of Species. New York: Random House.

Emerson, Richard M.
    1962 "Power-dependence relations." American Sociological Review, 27 (February): 31-41.

Ensor, R. C. K.
    1946. Some Reflections on Herbert Spencer's Doctrine that Progress Is Differentiation: The Herbert Spencer Lecture. London: Oxford University Press.

Epstein, Paul R., M.D.
    2006. "Avian Flu Is Already a World Pandemic," The New York Times, March 26.

Eyre, S. R.
>   1978 The Real Wealth of Nations. New York: St. Martin's Press.

Faris, Robert E. L.
>   1952 Social Psychology. New York: Ronald Press Co.

>   1955. Social Disorganization. New York: Ronald Press Co.

Frank, Thomas.
>   1995. "The Profit Value of Bad Family Values: Corporate America Has Found that Cultural Decay Is Very Good Business." Washington Post National Weekly Edition, June 19-25, p. 25.

Friedman, Thomas L.
>   2008 Hot, Flat, and Crowded: Why We Need a Green Revolution—and How It Can Renew America. New York: Farrar, Straus and Giroux.

Gallagher, Hugh Gregory.
>   1985. FDR's Splendid Deception. New York: Dodd, Mead & Company

Gelbspan, Ross
>   2004 Boiling Point: How Politicians, Big Oil and Coal, Journalists, and Activists Are Fueling the Climate Crisis—and What We Can Do to Avert Disaster. New York: Basic Books.

Gerges, Fawaz A.
>   2006. Journey of the Jihadist: Inside Muslim Militancy. Orlando, FL: Harcourt, Inc.

Gibbs, Jack P., and Walter T. Martin
>   1958. "Urbanization and Natural Resources." American Sociological Review, 23 (June): 266-277.

>   1962. "Urbanization, Technology, and the Division of Labor: International Patterns." American Sociological Review, 27 (Oct.): 667-677.

Gilbert, Michael.
>   2007 "Fort Lewis pilots, crew in copter crash identified." Tacoma, WA: The News Tribune, August 24.

Gillies, M.F.P., and N.A. Dodgson.
    1999. "Ball Catching: An Example of Psychologically-based Behavioural Animation." Paper presented to Eurographics UK 17th Annual Conference, Fitzwilliam College, Cambridge, UK, 13-15 April.

Glick, Thomas F., and David Kohn (eds.).
    1996. Darwin on Evolution: The Development of the Theory of Natural Selection. Indianapolis and Cambridge: Hackett Publishing Company, Inc.

Gold, Thomas, and Steven Soter
    1980 "The Deep-Earth-Gas Hypothesis." Scientific American, 242 (June): 130-137.

Goodell, Jeff
    2006. Big Coal: The Dirty Secret Behind America's Energy Future. Boston: Houghton Mifflin Company

Gorbachev, Mikhail S.
    1985 A Time for Peace. New York: Richardson & Steirman.

Gore, Al
    2006 An Inconvenient Truth: The Planetary Emergency of Global Warming and What We Can Do About It. New York: Rodale.

Gould, Stephen Jay.
    1999. Rocks of Ages: Science and Religion in the Fullness of Life. New York: Ballantine Publishing Group

    2002. The Structure of Evolutionary Theory. Cambridge, MA: The Belknap Press of Harvard University Press

Grist.
    2001. "Something Is Rotten in the State of Denmark: A skeptical look at The Skeptical Environmentalist." 12 Dec. On the web at: <www.grist.org/advice/books/2001/12/12/of/index.html>

Grodzins, Morton
    1949 Americans Betrayed: Politics and the Japanese Evacuation. Chicago: University of Chicago Press.

Gunther, John.
    1950. Roosevelt in Retrospect: A Profile in History. New York: Harper & Brothers, Publishers.

Haddock, S. H. D., C. W. Dunn, P. R. Pugh, C. E. Schnitzler
    2005 "Bioluminescent and red fluorescent lures in a deep-sea siphonophore." Science.309 (July 8).

Hafner, Katie, and Saritha Rai.
    2006. "Governments Tremble at Google's Bird's-Eye View." The New York Times, Dec. 20.

Hall, Bill.
    2006 "'The Deer' seems like subject fit for Hitchcock." Tacoma, WA: The News Tribune, April 8, p. B-5.

Hamilton, Robert A.
    2005. "If the Navy Leaves." The New York Times, June 5.

Hardin, Garrett
    1968 "The Tragedy of the Commons." Science, 162 (13 December): 1243-1248.

    1972. Exploring New Ethics for Survival: The Voyage of the Spaceship Beagle. New York: Viking Press.

Harris, Marvin
    1977. Cannibals and Kings: The Origins of Cultures. New York: Random House.

Haub, Carl.
    1995 "How Many People Have Ever Lived on Earth?" Population Today (February):4-5

Haupt, Lyanda Lynn.
    2006. Pilgrim on the Great Bird Continent: The Importance of Everything and Other Lessons from Darwin's Lost Notebooks. New York: Little, Brown and Company.

Hayakawa, S. I.
    1941. Language in Action. New York: Harcourt Brace and Company.

Hayes, Harold
    1981 "A Conversation with Garrett Hardin." The Atlantic Monthly, 247 (May): 60-70.

Heilbroner, Robert L.
    1974. An Inquiry into the Human Prospect. New York: W. W. Norton & Company, Inc.

Heinberg, Richard.
    2003. The Party's Over: Oil, War and the Fate of Industrial Societies. Gabriola Island, BC: New Society Publishers.

    2006 The Oil Depletion Protocol: A Plan to Avert Wars, Terrorism, and Economic Collapse. Gabriola, B.C.: New Society Publishers.

    2009 "The shape of the recovery." Energy Bulletin, January 7. <http://postcarbon.org/shape_recovery>

Henderson, Carter
    1978 The Inevitability of Petroleum Rationing in the United States. Princeton, NJ: Princeton Center for Alternative Futures, Inc.

Henderson, Nell
    2006. "Whither the Women?" Washington Post, July 7.

Hollin, John T., and Roger G. Barry
    1979. "Empirical and Theoretical Evidence Concerning the Response of the Earth's Ice and Snow Cover to a Global Temperature Increase." Environment International, 2: 461-473.

Holmes, Richard T., and Harry F. Recher.
    1986. "Search Tactics of Insectivorous Birds Foraging in an Australian Eucalypt Forest." The Auk, 103 (July):515-530.

Homer-Dixon, Thomas.
    2005. "Caught Up in Our Own Connections." The New York Times, August 13.

Howard, Fred.
    1987. Wilbur and Orville: A Biography of the Wright Brothers. New York: Alfred A, Knopf

Howe, John G.
>   2006 The End of Fossil Energy: And the Last Chance for Survival (3rd Edition). Waterford, ME: McIntire Publishing Services

Hsu, Spencer S.
>   2005. "Base Closing Plan's Legality Is Disputed by Sen. Warner." The New York Times, July 8.

Huang, Lucy Jen
>   1982 "Planned Fertility of One-couple/One-child Policy in the People's Republic of China." Journal of Marriage and the Family, 44 (August): 775-784.

Hubbert, M. King.
>   1969. "Energy Resources." Ch. 8 in Committee on Resources and Man, National Academy of Sciences, National Research Council, Resources and Man, pp. 157-242. San Francisco: W. H. Freeman.

Huber, Peter W., and Mark P. Mills.
>   2005. The Bottomless Well: The Twilight of Fuel, the Virtue of Waste, and Why We Will Never Run Out of Energy. New York: Basic Books.

Hutchinson, G. E.
>   1965. The Ecological Theater and the Evolutionary Play. New Haven, CT: Yale University Press.

Iacocca, Lee (with William Novak).
>   1984. Iacocca: An Autobiography. New York: Bantam Books.

Johnson, Carrie.
>   2005. "WorldCom's Ebbers Sentenced to 25 Years." Washington Post, July 13.

Johnson, Kirk.
>   2006. "Clogged Rockies Highway Divides Coloradans." The New York Times, January 25.

Johnson, Michael P., and Peter H. Raven.
>   1970. "Natural Regulation of Plant Species Diversity." Evolutionary Biology, 4:127-162.

Jones, Robert Alun.
    1994a. "The Positive Science of Ethics in France: German Influences on De la Division du Travail Social." Sociological Forum, 9 (March):27-57.

    1994b "Ambivalent Cartesians: Durkheim, Montesquieu, and Method." American Journal of Sociology, 100 (July):1-39.

Kay, Jane
    2007. "Birthing bears head for land as Arctic ice gets scarcer." San Francisco Chronicle, July 13.

Kessler, David.
    2001. A Question of Intent: A Great American Battle with a Deadly Industry. New York: Public Affairs

Keynes, Randal.
    2002. Darwin, His Daughter and Human Evolution. New York: Riverhead Books

King, Sir David
    2007 Interview, Science, 21 Dec., pp. 1862-1863

King, Michael.
    2003. The Penguin History of New Zealand. Auckland: Penguin Books (NZ) Ltd.

Korzybski, Alfred.
    1948. Science and Sanity: An introduction to non-Aristotelian systems and general semantics. 3rd ed. Lakeville, CT: International Non-Aristotelian Library Pub. Co.

    1950. Manhood of Humanity (Second ed.). Lakeville, CT: Institute of General Semantics, 1950.

Krisher, Tom.
    2006. "U.S. Automakers Turn to Dems for Help." Washington Post, Nov. 8.

Kurtz, Howard.
    2006. "Mine Disaster's Terrible Irony: A Failure to Look Deeper." Washington Post, Jan. 9, p. C-1.

Laurmann, J. A.
　　　1979 "Climate Change from Fossil Fuel Generated CO2 and Energy Policy." Environment International, 2: 461-473.

Leakey, Richard E.
　　　1981 The Making of Mankind. New York: E. P. Dutton

Leakey, Richard, and Roger Lewin
　　　1992. Origins Reconsidered: In Search of What Makes Us Human. New York: Doubleday

Ledec, George
　　　1985 "The Political Economy of Tropical Deforestation." pp. 179-226 in H. Jeffrey Leonard (ed.), Divesting Nature's Capital: The Political Economy of Environmental Abuse in the Third World. New York: Holmes & Meier.

Lee, Don.
　　　2006. "China's Chopstick Tax Seems Dim to Some." The Los Angeles Times, March 24.

Lenski, Gerhard, and Jean Lenski
　　　1982 Human Societies: An Introduction to Macrosociology (4th ed.). New York: McGraw-Hill.

Leonard, H. Jeffrey
　　　1984 Are Environmental Regulations Driving U.S. Industry Overseas? Washington, DC: The Conservation Foundation.

Lieberman, Philip.
　　　1998. Eve Spoke: Human Language and Human Evolution. New York: W. W. Norton.

Lindgren, Hugo.
　　　2002. "THE YEAR IN IDEAS; Know-Nothing C.E.O., The." The New York Times. December 15.

Lipset, Seymour Martin, and William Schneider.
　　　1983 The Confidence Gap: Business, Labor, and Government in the Public Mind. New York: The Free Press.

Lloyd, E. M. H.
    1943 "The Food Situation in a World at War: In the United Kingdom."
    Annals of the American Academy of Political and Social Science,
    225 (January): 83-85.

Lo, Clarence H. Y.
    2000 "Alienation." Encyclopedia of Sociology (Second edition) vol. 1,
    pp. 99-106. Edgar F. Borgatta and Rhonda J. V. Montgomery
    (eds.). New York: Macmillan Reference USA.

Lomborg, Bjørn.
    2001. The Skeptical Environmentalist: Measuring the Real State of the
    World. Cambridge, UK, and New York: Cambridge University
    Press.

Love, Donald M.
    1956. Henry Churchill King, of Oberlin. New Haven, CT: Published
    for Oberlin College by Yale University Press.

Lovelock, James.
    2006. The Revenge of Gaia: Earth's Climate in Crisis and the Fate of
    Humanity. New York: Basic Books

Lovins, Amory B.
    2006 "Leaving Appalachia Right Side Up ... At a Profit." Orion
    (January-February): <http://www.orionmagazine.org/index.
    php/articles/article/262/>

Lovins, Amory B., and L. Hunter Lovins
    1982 Brittle Power: Energy Strategy for National Security. Andover,
    MA: Brick House Publishing Co.
Lozano, Juan A.
    2006. "Enron's Fastow Assigned to La. Prison." Washington Post (The
    Associated Press), November 10.

Lumsden, Charles J., and Edward O. Wilson
    1983 Promethean Fire: Reflections on the Origin of Mind. Cambridge,
    MA: Harvard University Press.

Lundberg, George A.
    1947. Can Science Save Us? New York: Longmans, Green and Co.

Lynd, Robert S. & Lynd, Helen Merrell.
    1929. Middletown. London: Constable.

MacArthur, Robert.
    1969. "Species Packing, and What Interspecies Competition
        Minimizes." Proceedings of the National Academy of Sciences,
        U.S. 64 (Sept.-Dec.): 1369-1371.

    1970. "Species Packing and Competitive Equilibrium for Many
        Species." Theoretical Population Biology, 1:1-11.

Machiavelli, Nicolò
    1997 [1532] The Prince (Translated and edited by Angelo M. Codevilla).
        New Haven: Yale University Press.

MacKinnon, John
    1978 The Ape Within Us. New York: Holt, Rinehart, and Winston.

Maddox, John R.
    1972. The Doomsday Syndrome. New York: McGraw-Hill.

Malinowski, Bronislaw
    1947 Freedom and Civilization. New York: Roy Publishers.

Markowitz, Gerald, and David Rosner.
    2002. Deceit and Denial: The Deadly Politics of Industrial Pollution.
        Berkeley and New York: University of California Press and The
        Milbank Memorial Fund

Marsden, Sullivan, Jr.
    1986 Letter. Christian Science Monitor, February 12, p. 15.

Marshall, Kathryn.
    1987. In the Combat Zone: An Oral History of American Women in
        Vietnam, 1966-1975. Boston: Little, Brown and Company.

Martin, Andrew, and Patrice Petro (eds.).
    2006. Rethinking global security: media, popular culture, and the
        "War on terror." New Brunswick, N.J.: Rutgers University Press,
        2006.

Martin, Mike W (Ed.).
>   1985 Self-Deception and Self-Understanding: New Essays in Philosophy and Psychology. Lawrence: University Press of Kansas.

Mauss, Armand.
>   1975. Social problems as social movements. Philadelphia : Lippincott.

Maxtone-Graham, John.
>   1972. The Only Way to Cross. New York: Barnes & Noble Books.

McBeath, Michael K., Dennis M. Shaffer, Mary K. Kaiser.
>   1995 "How Baseball Outfielders Determine Where to Run to Catch Fly Balls." Science, 28 April 1995, Volume 268, pp. 569-573.

McCrone, John.
>   1991 The Ape that Spoke: Language and the Evolution of the Human Mind. New York : William Morrow and Co.

McGrail, Sean
>   2002. Boats of the World: From the Stone Age to Medieval Times. New York: Oxford University Press

McHugh, David.
>   2006. "Bird Flu Found in Stone Marten in Germany." Washington Post. March 10.

McKibben, Bill
>   1992 The Age of Missing Information. New York: Random House.

McKie, Robin.
>   2000 Dawn of Man: The Story of Human Evolution. New York: Dorling Kindersley Publishing, Inc.

McNeil, Donald G. Jr.
>   2006 "At the U.N.: This Virus Has an Expert 'Quite Scared'." The New York Times, March 28.

McNeal, Gregory S.
>   2005 "The Terrorist and the Grid." The New York Times, August 13.

Merton, Robert K.
    1934 "Durkheim's Division of Labor in Society." American Journal of Sociology, 40 (November):316-328.

Merton, Robert K., and Elinor Barber.
    2004 The Travels and Adventures of Serendipity: A Study in Sociological Semantics and the Sociology of Science. Princeton, NJ: Princeton University Press

Meyers, Robin.
    2006 Why the Christian Right Is Wrong: A Minister's Manifesto for Taking Back Your Faith, Your Flag, Your Future. New York: Jossey-Bass.

Milbrath, Lester W.
    1989 Envisioning a Sustainable Society: Learning Our Way Out. Albany: State University of New York Press.

Mills, C. Wright.
    1958 The Causes of World War Three. New York: Simon and Schuster.

Milward, Alan S.
    1977 War, Economy and Society. London: Allen Lane.

Mohajer, Shaya Tayefe
    2007. "Powerball Win: Fantasy or Nightmare?" Washington Post, Sept. 14.

Molander, Roger C., David A. Mussington, and Peter A. Wilson.
    1998 Cyberpayments and Money Laundering: Problems and Promise. Santa Monica, CA: RAND.

Moorcraft, Colin.
    1973 Must the Seas Die? Boston: Gambit.

Morell, Virginia
    2007 "Can the Wild Tiger Survive?" Science, 317 (7 September): 1312-1314.

Morison, Samuel Eliot.
    1965 The Oxford History of the American People. New York: Oxford
         University Press

Morris, Charles R.
    2006 "Freakoutonomics." The New York Times, June 2

Morris-Cotterill, Nigel.
    2001 "Money Laundering." Foreign Policy, May/June.

Moulson, Geir
    2007 "Pelosi: Climate Change Is a Reality." Washington Post, May 28.

Mountfort, Guy
    1981 Saving the Tiger. New York: Viking Press.

Mulloy, William.
    1974 "Contemplate the Navel of the World." Americas, 26 (April):
         25-33.

Nierenberg, Danielle.
    2002 Correcting Gender Myopia: Gender Equity, Women's Welfare,
         and the Environment. Washington, DC: Worldwatch Institute.

Nofz, Michael P.
    1983 "Carrying Capacity and U.S. Agriculture." The Rural Sociologist,
         3 (September): 303-311.

Norris, Floyd.
    2002 "For Chief, $200 Million Wasn't Quite Enough Cash." The New
         York Times, January 22.

Odum, Eugene P.
    1989 Ecology: And Our Endangered Life-Support Systems.
         Sunderland, MA: Sinauer Associates, Inc.

Odum, Howard T.
    1971 Environment, Power, and Society. New York: Wiley—
         Interscience.

Oestreicher, David.
   1973 "Growing, crowding population hiking earth's thermostat." Detroit Free Press, Tuesday, June 12, p. 12-A.

Ogburn, William F.
   1966 Social Change: With Respect to Cultural and Original Nature. New York: Dell Publishing Co., Inc. [reprint of the 1922 original edition, B. W. Huebsch]

Oleksyn, Veronika
   2006 "Some EU Nations Offer Benefits for Births." Washington Post, July 17.

Olzak, Susan, Suzanne Shanahan, and Elizabeth West.
   1994 "School Desegregation, Interracial Exposure, and Antibusing Activity in Contemporary Urban America." American Journal of Sociology, 100 (July):196-241.

Ophuls, William, and A. Stephen Boyan, Jr.
   1992 Ecology and the Politics of Scarcity Revisited: The Unraveling of the American Dream. New York: W. H. Freeman and Company.

Oppenheimer, Michael
   1986 "Will the Planet Remain Habitable?" The New York Times, June 30, p.A19.

Organic Farming Research Foundation, Santa Cruz, CA.
   1999. "Final Project Report: Bird and arthropod predation of codling moth in apple orchards." February. JoAnn Baumgartner, Principal Investigator

Ostheimer, John M., and Leonard G. Ritt
   1982 "Abundance and American Democracy: A Test of Dire Predictions." Journal of Politics, 44 (May): 366-387.

Owen, Denis
   1980 Camouflage and Mimcry. Chicago: The University of Chicago Press.

Ozinga, James R.
    1985 The Prodigal Human. Jefferson, NC: McFarland & Company,
        Publishers.

Pacala, S. W., and Jonathan Roughgarden.
    1982 "The Evolution of Resource Partitioning in a Multidimensional
        Resource Space." Theoretical Population Biology, 22:127-145.

Paletz, David L., and Alex P. Schmid (eds.).
    1992 Terrorism and the Media. Newbury Park, Calif.: Sage.

Pan, Philip P.
    2006 "The Click That Broke a Government's Grip." Washington Post,
        Feb. 19, p. A-1.

Parker, Kathleen.
    2005 "They Shoot Women, Don't They?" Orlando Sentinel. June 29.

Parsons, Jack
    1971 Population versus Liberty. London: Pemberton Books.

Pérez-Peña, Richard.
    2005 "Strike Inflicts Broad Economic Pain." The New York Times,
        December 21.

Perrin, Robert G.
    1976 "Herbert Spencer's Four Theories of Social Evolution." American
        Journal of Sociology, 81 (May):1339-1359.

Petersen, Melody
    2008 Our Daily Meds: How the Pharmaceutical Companies
        Transformed Themselves into Slick Marketing Machines and
        Hooked the Nation on Prescription Drugs. New York: Sarah
        Crichton Books (Farrar, Straus and Giroux)

Peterson, Ivars.
    2002 "Catching Flies." Science News Online, 162 (Oct. 12):15

Petitto, L. A., and Marentette, P.
    1991 "Babbling in the manual mode: Evidence for the ontogeny of
        language." Science, 251:1483-96.

Piller, Charles.
2006 "As fears of flu thrive, so do pills and promises." The Los Angeles Times, Jan 2.

Pollan, Michael
2001 The Botany of Desire: A Plant's Eye View of the World. New York: Random House

Popper, Karl.
1968 "Plato." vol. 12, pp. 159-164 in David L. Sills (ed.), International Encyclopedia of the Social Sciences. New York: Macmillan and The Free Press.

Porter, Eduardo.
2006 "Detroit Can't Count on History Repeating." The New York Times, April 14.

Portes, Alejandro, and Julia Sensenbrenner.
1993 "Embeddedness and Immigration: Notes on the Social Determinants of Economic Action." American Journal of Sociology, 98 (May):1320-1350.

Poston, Dudley L., and Mei-Yu Yu
1985 "Quality of Life, Intellectual Development and Behavioural Characteristics of Single Children in China: Evidence from a 1980 Survey in Changsha, Hunan Province." Journal of Biosocial Science, 17 (April): 127-136.

Potter, David.
1954 People of Plenty: Economic Abundance and the American Character. Chicago: University of Chicago Press.

Powell, Jeffrey R., and Charles E. Taylor.
1979 "Genetic Variation in Ecologically Diverse Environments." American Scientist, 67 (Sept./Oct.):590-596.

Price, Peter W.
1984 "Alternative Paradigms in Community Ecology." pp. 353-383 in Peter W. Price, C. N. Slobodchikoff, and William S. Gaud (eds.), A New Ecology: Novel Approaches to Interactive Systems. New York: John Wiley & Sons.

Pringle, Peter.
    1998 Cornered: Big Tobacco at the Bar of Justice. New York: Henry
        Holt and Company

Quammen, David.
    2006 The Reluctant Mr. Darwin. New York: W. W. Norton &
        Company.

Raasch, Chuck.
    1995 "Who's Sour Now? Public Tops Press in Cynicism, Poll Finds."
        (Gannett News Service) The News Tribune, Tacoma, WA. May
        22, p. B6.

Reader, John.
    1981 Missing Links: The Hunt for Earliest Man. Boston: Little, Brown
        and Company.

Reece, Erik
    2006 "Moving Mountains: The battle for justice comes to the coal
        fields of Appalachia." Orion (January-February): <http://www.
        orionmagazine.org/index.php/articles/article/166/>

Reich, Charles.
    1970 The Greening of America. New York: Random House.

Reich, Robert B.
    1991 The Work of Nations. New York: Alfred A. Knopf.

Reuters.
    2002 "Enron CEO: Rock-Star CEOs led to Scandals." The New York
        Times, October 15.

Rickleffs, R. E.
    1979 Ecology, 2nd ed. New York: Chiron Press.

Reining, Priscilla
    1979 Challenging Desertification in West Africa: Insights from Landsat
        into Carrying Capacity, Cultivation, and Settlement Sites in
        Upper Volta and Niger. Athens, OH: Ohio University Center
        for International Studies.

Roberts, Paul.
    2004 The End of Oil: On the Edge of a Perilous New World. Boston: Houghton Mifflin Company

Robinson, Eugene.
    2006 "Nation of Fear." Washington Post, May 16.

Rosecrance, Richard.
    1986 The Rise of the Trading State. New York: Basic Books, Inc.

Ross, Edward Alsworth.
    1907 Sin and Society: An Analysis of Latter-Day Iniquity. Boston: Houghton Mifflin [reprinted 1973 by Harper Torchbooks].

Rostow, Walt W.
    1963 The Economics of Take-Off into Sustained Growth. New York: St. Martin's Press.

Roughgarden, Jonathan.
    1976 "Resource Partitioning Among Competing Species—A Coevolutionary Approach." Theoretical Population Biology. 9:388-424.

Rumsfeld, Donald H.
    2006 "War in the Information Age." The Los Angeles Times, Feb. 23.

Russo, Thomas A.
    2002 Mechanical Typewriters: Their History, Value, and Legacy. Atglen, PA: Schiffer Publishing Ltd.

Samuelson, Robert J.
    2006 "Delinquency Of the CEOs." Washington Post, July 13, p. A23.

Saywell, Shelley
    1985 Women in War. New York: Viking.

Scheuer, Jeffrey.
    1999 The Sound Bite Society: Television and the American Mind. New York: Four Walls Eight Windows.

Schlesinger, Arthur M., Jr.
  1992 The Disuniting of America: Reflections on a Multicultural Society. New York: W. W. Norton.

Schoener, T. W.
  1974 "Resource Partitioning in Ecological Communities." Science. 185 (5 July): 27-39.

  1989 "The Ecological Niche." pp. 79-113 in J. M. Cherrett (ed.), Ecological Concepts: The Contribution of Ecology to an Understanding of the Natural World. Oxford, UK: Blackwell Scientific Publications.

Schnaiberg, Allan.
  1980 The Environment: From Surplus to Scarcity. New York: Oxford University Press.

Schnitzler, Hans-Ulrich, Cynthia F. Moss, and Annette Denzinger.
  2003 "From spatial orientation to food acquisition in echolocating bats." TRENDS in Ecology and Evolution, 18 (August): 386-394

Seago, David.
  2006 "EDITORIAL NOTE: Islam cartoons won't play in Tacoma." Tacoma, WA: The News Tribune, Feb. 8, p. B-6.

Sears, Paul B.
  1935 Deserts on the March. Norman, OK: University of Oklahoma Press.

Sellin, Thorsten
  1938 Culture Conflict and Crime. New York: Social Science Research Council.

Shabecoff, Philip
  1986 "Swifter Warming of Globe Foreseen: Climate and Health Imperiled by Atmospheric Changes, Scientists Tell Senate." The New York Times, June 11.

Siegel, Marc.
  2006 "The cost of bird flu hysteria." Boston Globe, March 17.

Simmel, Georg
1978 The Philosophy of Money (transl. by Tom Bottomore and David Frisby). London and Boston: Routledge & Kegan Paul.

Simon, Julian L., and Herman Kahn.
1984 The Resourceful Earth: A Response to'Global 2000.' New York: Basil Blackwell.

Simpson, Evelyn, Helen Gardner, and Timothy Healy (eds.)
1967 John Donne: Selected Prose. Oxford, UK: Clarendon Press.

Singh, Arjan
1984 Tiger! Tiger! London: Jonathan Cape.

Skelton, George.
2006. "Hetch Hetchy Restoration Is a Pricey Pipe Dream." The Los Angeles Times, Nov. 30.

Smith, Adam.
1965 [1776] An Inquiry into the Nature and Causes of the Wealth of Nations. Edited, with an Introduction, Notes, Marginal Summary and Enlarged Index by Edwin Cannan. With an Introduction by Max Lerner. New York: The Modern Library.

Smith, Dennis.
1983 Barrington Moore: Violence, Morality and Political Change. London: Macmillan.

Sorokin, Pitirim A.
1957 Social and Cultural Dynamics (1 vol. ed.). Boston: Porter Sargent.

Spencer, Herbert.
1958 [1862] First Principles. New York: The De Witt Revolving Fund, Inc.

1961[1873] The Study of Sociology (Introduction by Talcott Parsons). Ann Arbor: University of Michigan Press.

Stanley, Manfred.
1978 "Dignity versus survival? Reflections on the moral philosophy of social order." pp. 197-234 in R. H. Brown and S. M. Lyman

(eds.), Structure, Consciousness, and History. Cambridge, UK: Cambridge University Press.

Starke, Linda (ed.).
　　2007 State of the World 2007: Our Urban Future. New York: W. W. Norton & Company.

Stein, Benjamin J.
　　1992 A License to Steal. New York: Simon & Schuster.

Stein, R.C.
　　1958 "The behavioral, ecological and morphological characteristics of two populations of the Alder Flycatcher, Empidonax trailli (Audubon)." New York State Museum of Science Service Bulletin No. 371.

Stokoe, William C.
　　2001 Language in Hand: Why Sign Came Before Speech. Washington, DC: Gallaudet University Press

Stoll, Clifford
　　1989 The Cuckoo's Egg: Tracking a Spy through the Maze of Computer Espionage. New York: Doubleday

Stringer, Christopher, and Robin McKie.
　　1996 African Exodus: The Origins of Modern Humanity. New York: Henry Holt and Company.

Strother, Robert S., and Eugene H. Methvin.
　　1975 "Terrorism on the Rampage." Reader's Digest, November.

Strouse, Jean.
　　1999 Morgan: American Financier. New York: Random House.

Sumner, William Graham
　　1896 "Earth Hunger or the Philosophy of Land Grabbing." pp. 31-64 in A. G. Keller (ed.), Earth Hunger and Other Essays. New Haven, CT: Yale University Press.

Sutherland, Edwin H., and Donald R. Cressey
　　1960 Principles of Criminology. Philadelphia: Lippincott

Szep, Jason.
　　2006 "US facing wave of murders and gun violence." Washington Post, August 20.

Tattersall, Ian
　　1998 Becoming Human: Evolution and Human Uniqueness. New York: Harcourt Brace

Thomas, Ken.
　　2006 "Bush to Meet With U.S. Automakers. Washington Post, Nov. 14.

Thomas, William I.
　　1923 The Unadjusted Girl. Boston: Little, Brown.

Thompson, Jack.
　　2006 "Violent video games feed unhealthy ideas to young kids." The News Tribune, Tacoma, WA, Jan. 8, Section: Insight, p. 4.

Thompson, J. N.
　　1982 Interaction and Coevolution. New York: Wiley.

Thomson, James T.
　　1985 "The Politics of Desertification in Marginal Environments: The Sahelian Case." pp. 227-262 in H. Jeffrey Leonard (ed.), Divesting Nature's Capital: The Political Economy of Environmental Abuse in the Third World. New York: Holmes & Meier.

Tuchman, Barbara
　　1981 Practicing History: Selected Essays. New York: Alfred A. Knopf

Tucker, Eric.
　　2005 "Deadly Attack in Fallujah Puts Focus on Women in Military." Boston Globe, July 3.

Twain, Mark [Samuel Clemens]
　　1889 A Connecticut Yankee in King Arthur's Court. New York: Charles L. Webster & Co.

Urbina, Ian.
　　2006 "In Online Mourning, Don't Speak Ill of the Dead." The New York Times, November 5.

U.S. Code, Title 18, Part I, Chapter 95 § 1956 Laundering of monetary instruments

Valentine, James W.
1971 "Resource Supply and Species Diversity Patterns." Lethaia, 4:51-61.

Van den Berghe, Pierre L.
1973 Age and Sex in Human Societies: A Biosocial Perspective. Belmont, CA: Wadsworth Pub. Co.

Vayda, Andrew P.
1974 "Warfare in Ecological Perspective." Annual Review of Ecology and Systematics, 5: 183-193.

Vogt, William.
1948 Road to Survival. New York: William Sloan Associates.

Wackernagel, Mathis, and William E. Rees
1996 Our Ecological Footprint: Reducing Human Impact on the Earth. Gabriola Island, BC: New Society Publishers.

Wald, Matthew L.
2006 "F.A.A. Finds More Errors on Runways" The New York Times, Nov. 3.

Wallace, Bruce.
1972 People, Their Needs, Environment, Ecology: Essays in Social Biology. Englewood Cliffs, NJ: Prentice-Hall.

Webb, Walter Prescott.
1952 The Great Frontier. Boston: Houghton Mifflin.

Wekesser, Carol, and Matthew Polesetsky (eds.)
1991 Women in the Military. San Diego, CA: Greenhaven Press.

Wells, Joseph T.
2003 "Money Laundering: Ring Around the White Collar." Journal of Accountancy, June.

Whelpton, P. K.
  1939 "Population Policy for the United States." Journal of Heredity, 30 (September), 401-406.

Wickler, Wolfgang
  1968 Mimicry: In Plants and Animals (Transl. from German by R. D. Martin). New York: McGraw-Hill.

Wilford, John Noble.
  1990 "Mastermind of Piltdown Hoax Unmasked?" The New York Times, June 5.

  2003 "How the Wright Brothers Did What No One Else Could." The New York Times, Dec. 9.

Willis, David K.
  1985a "Chinese Appear Confident: Only Children Won't Be Spoiled Children." Christian Science Monitor, December 16, p. 9.

  1985b "Chinese Grandmas Sing the Party Line." Christian Science Monitor, December 17, p. 10.

Wilson, James Q.
  1968 "Why We Are Having a Wave of Violence." The New York Times Magazine, May 19

Winchester, Simon.
  1998 The Professor and the Madman: A Tale of Murder, Insanity, and the Making of the Oxford English Dictionary. New York: HarperCollins Publishers, Inc.

Wittfogel, Karl August
  1968 "The Theory of Oriental Society." pp. 179-198 in Morton H. Fried (ed.), Readings in Anthropology (2nd ed.), vol. 2: Cultural Anthropology. New York: Thomas Y. Crowell Company.

World Bank
  1980 World Tabled (2nd ed.). Baltimore, MD: Johns Hopkins University Press.

World Resources Institute, and International Institute for Environment and Development
  1986 World Resources 1986: An Assessment of the Resource Base that Supports the Global Economy. New York: Basic Books, Inc.

Yang, Charles.
  2006 The Infinite Gift: How Children Learn and Unlearn the Languages of the World. New York: Scribner

Zernike, Kate.
  2006 "Violent Crime Rising Sharply in Some Cities." The New York Times, Feb. 12.

Zimmer, Carl
  2008 "What Is a Species?" Scientific American, 298 (June):72-79.

LaVergne, TN USA
26 November 2010
206392LV00008B/28/P